Cognitive Systems Monographs

Volume 6

Editors: Rüdiger Dillmann · Yoshihiko Nakamura · Stefan Schaal · David Vernon

Helge Ritter, Gerhard Sagerer,
Rüdiger Dillmann, and Martin Buss (Eds.)

Human Centered Robot Systems

Cognition, Interaction, Technology

Springer

Rüdiger Dillmann, University of Karlsruhe, Faculty of Informatics, Institute of Anthropomatics, Humanoids and Intelligence Systems Laboratories, Kaiserstr. 12, 76131 Karlsruhe, Germany

Yoshihiko Nakamura, Tokyo University Fac. Engineering, Dept. Mechano-Informatics, 7-3-1 Hongo, Bukyo-ku Tokyo, 113-8656, Japan

Stefan Schaal, University of Southern California, Department Computer Science, Computational Learning & Motor Control Lab., Los Angeles, CA 90089-2905, USA

David Vernon, Khalifa University Department of Computer Engineering, PO Box 573, Sharjah, United Arab Emirates

Authors

Professor Dr. Helge Ritter
Universität Bielefeld
Technische Fakultät
AG Neuroinformatik
Universitätsstr. 25
33615 Bielefeld

Professor Dr.-Ing. Gerhard Sagerer
Universität Bielefeld
Technische Fakultät
AG Angewandte Informatik
Universitätsstr. 25
33615 Bielefeld

Prof. Dr.-Ing. Rüdiger Dillmann
Uni Karlsruhe
Technologiefabrik
Haid-und-Neu-Strasse 7
D-76131 Karlsruhe

Prof. Dr.-Ing. Martin Buss
Institute of Automatic Control
Engineering (LSR)
Technische Universität München
Arcisstraße 21
80290 München
Germany

ISBN 978-3-642-10402-2 e-ISBN 978-3-642-10403-9

DOI 10.1007/978-3-642-10403-9

Cognitive Systems Monographs ISSN 1867-4925

Library of Congress Control Number: 2009938822

Typeset & Cover Design: Scientific Publishing Services Pvt. Ltd., Chennai, India.

Printed in acid-free paper

5 4 3 2 1 0

springer.com

Preface

Making robots interact with humans in a useful manner can be approached in two complementary directions: taking advantage of the flexibility of human cognition to simplify the interaction for the robot; or conversely, trying to add as much flexibility and "cognition" into the robot system to simplify the interaction for the human.

Limitations of today's technology and of our understanding of how to achieve highly flexible "cognitive" robot behavior has led to the development and deployment of robot systems almost exclusively in the first direction. However, in recent years we have seen that we are nearing the point where we can build robots that can interact with people in more flexible ways, integrating significant abilities of perception, navigation, communication, situation awareness, decision making and learning.

This development has reshaped the landscape of robotics research in a significant way and has created a new focus with fascinating perspectives from both the perspective of scientific research and the perspective of practical applications: the realization of Human Centered Robot Systems.

These systems must be able to interact with humans in natural ways that shift the burden of adaptation to the machine and not the human. Therefore, their interaction and communication patterns must comply with the level of human abstraction and with the ways of human cognition. As a consequence, their realization cannot be viewed as an engineering task alone, but requires the integration of insights from a considerable spectrum of other disciplines that try to nurture our understanding of the "Gestalt" of cognition and its underlying mechanisms.

The present volume collects a set of prominent papers presented during a two-day conference on *Human Centered Robotic Systems* held in Bielefeld during Nov. 19-20, 2009. The goal of this conference – with predecessors in Karlsruhe 2002 and in Munich 2006 – was to bring together researchers in the area focusing on the shared goal of understanding, modeling and implementing cognitive interaction from the perspective of robotics, computer science, psychology, linguistics, and biology in order to advance interdisciplinary dialogue and to enhance our knowledge towards the realization of Human Centered Robotic Systems in both the mentioned complementary directions.

The papers in this volume offer a survey of recent approaches, the state-of-the-art, and advances in this interdisciplinary field. They hopefully can provide both practitioners and scientists with an up-to-date introduction into a very dynamic field, which has not yet attained text book status, but which is rich with fascinating scientific challenges. Methods and solutions in this area are very likely to produce a significant impact on our future lives.

As editors, we would like to thank the authors for their work and the members of the program committee for the reviews of the submitted manuscripts. We would also like to thank Dr. Katharina Tluk von Toschanowitz for the coordination of the compilation of this volume.

Bielefeld
September 2009

Helge Ritter
Gerhard Sagerer
Martin Buss
Rüdiger Dillmann

Contents

System Integration Supporting Evolutionary Development and Design

Thorsten P. Spexard and Marc Hanheide

Abstract. With robotic systems entering our daily life, they have to become more flexible and subsuming a multitude of abilities in one single integrated system. Subsequently an increased extensibility of the robots' system architectures is needed. The goal is to facilitate a long-time evolution of the integrated system in-line with the scientific progress on the algorithmic level. In this paper we present an approach developed for an event-driven robot architecture, focussing on the coordination and interplay of new abilities and components. Appropriate timing, sequencing strategies, execution guaranties, and process flow synchronization are taken into account to allow appropriate arbitration and interaction between components as well as between the integrated system and the user. The presented approach features dynamic reconfiguration and global coordination based on simple production rules. These are applied first time in conjunction with flexible representations in global memory spaces and an event-driven architecture. As a result a highly adaptive robot control compared to alternative approaches is achieved, allowing system modification during runtime even within complex interactive human-robot scenarios.

1 Introduction

Robots are developing from very specific tools towards generic and flexible assistants or even companions. With the applicable fields getting broader and the abilities of the robots increasing, it is simply not feasible to construct a robot system from scratch. In contrary, a robotic system should evolve and successively incorporate new abilities and functions as soon as they are available and needed. However, robotic research is still far away from its maturity and still highly dynamic. Thus, an integration approach that paces up with the fast development in robotics, needs to

Thorsten P. Spexard · Marc Hanheide
Applied Informatics, Bielefeld University, Germany
e-mail: {tspexard,mhanheid}@techfak.uni-bielefeld.de

H. Ritter et al. (Eds.): Human Centered Robot Systems, COSMOS 6, pp. 1–9.
springerlink.com

embrace change and be flexible in its particular integration strategy. We present our approach focussing on the techniques to create robotic systems which fulfill these requirements resulting in an *Evolutionary System Integration Architecture ESIA*, which evolves with new demands. Our general focus thereby lies on *interactive* robots, that feature sets of different behaviors and abilities. Though it is self-evident that for any successful software system a detailed architecture has to be designed in advance, reality shows that during software integration often unforeseen challenges occur and adaptations have to be made quickly. Thus ESIA supports simple and fast, well-structured mechanisms for system modifications without creating an inscrutable kludge. Since in the scientific community larger-scale robotic systems are constructed in joint efforts, like in the European Integrated Projects CoSy [5] and Cogniron [12]), system integration can also pose a social challenge. Thus, aspects such as "informational value" and "familiarization time" are relevant as well. This integration of previous independently developed robot functions into one system in rather short time frames can be seen as a typical problem of legacy integration.

To accept this challenge our evolving integration approach makes use of a software architecture proposed by Wrede as information-driven integration (IDI) [13]. It combines benefits of event-driven architecture and space-based collaboration applying flexible representation of information using the XML data set. Though the IDI approach indeed serves as a foundation to evolutionary system integration with respect to interface changes and flexible orchestration, it does not strive to provide a comprehensive solution to the question "How to coordinate components and how to design their interplay?". In our presented approach we strive for a generic solution to this challenge, which shall (i) maintain a highly decoupled system architecture while (ii) still allowing flexible control and arbitration necessary for the ever changing requirements of different robotic functions. A central organizer is used to coordinate the separate module operations to the desired system behavior regarding both, serialization, and arbitration of the different module actions. In contrast to usual system controllers or planners the organizer is not directly connected with the system components but simply modifies the memory content based on rules in such a way that the desired behavior emerges.

2 Related Work

Derived from classical AI a rule-production-system (RPS) uses a set of rules, a working memory, and an interpreter to solve tasks. Rules consist of requirements which have to be fulfilled so that certain actions take place. For the evaluation whether a condition is satisfied the data within the working memory is used. The robot Caesar is using RPS [11] by a rule based controller, set on the top of a modular system with multiple layers of different abstraction levels. In contrast to ESIA only on the bottom level standard programming language is used for sensor and actuator drivers. For all remaining modules Erlang is used as demonstrator specific runtime environment. The module information is read once on start up and is based on fixed interface descriptions. Thus algorithms created in different and more public programming languages have to be re-implemented especially on the basis of the system specific interface needed by the controller. We propose the support of various widespread programming languages like C(++) or Java in combination with a generic underlying data model, e.g., XML representation.

Another aspect is the flexibility of rule-production-systems during runtime. Though easily adaptable when the system is shut down Ishida discusses in [7] the opportunity in adapting the behavior during runtime. A meta programming language is used to dynamically change the rules themselves while the system works. We adapted these ideas by assigning the rule evaluation to separate threads with one thread per rule and developing a way to dynamically change rules during runtime. Instead using another programming language, or more general another representation, we propose the idea of joining the representation of behavior rules and the data they apply to. Furthermore the rules are no longer separated from the working memory and set on the top of the system, but are memory content which is equally treated as, e.g., processed sensor data.

The rule definition complies with the Event-Condition-Action model introduced by [4]. To satisfy the requirement of a rule, so that the associated action takes place, not only a certain event has to take place, but additionally the current system situation is considered by an additional condition which has to be complied with. Using the system state in combination with a recent event, involves the handling of temporal dependencies since the events leading to the desired system state had to take place prior to the event finally triggering the action execution of a rule. Although the importance of temporal dependencies is described in [10], temporal aspects in evaluating rules are only referred to in terms of a given global execution time which has to be achieved. The question, how to deal with, e.g., temporal concurrency in parallel systems is left open.

3 Creating Systems for HRI

We implemented the presented approach on two different robotic demonstrators focussing especially on extended abilities for advanced human-robot interaction. One demonstrator is BIRON as depicted in Fig. 1(a), which is used in a Home-Tour scenario [9]. In this scenario naive users have to guide the robot around in a real world flat to familiarize the robot with a new vicinity. The users receive only a brief introduction to the system and have to intuitively interact with it by naturally spoken dialog and gestures. The results taken from the user interactions have to be simply integrated into the existing robotic system for continuous improvement. Additionally to system modifications on a single type of demonstrator in one scenario ESIA adapts to alternate demonstrators and scenarios. In a second scenario we use the anthropomorphic robot BARTHOC (see Fig. 1(b)) in a receptionists setting [2]. In this setting robot and human oppose each other at a table with a map of the surrounding lain between them. The interaction partner refers to the map in a natural way and the

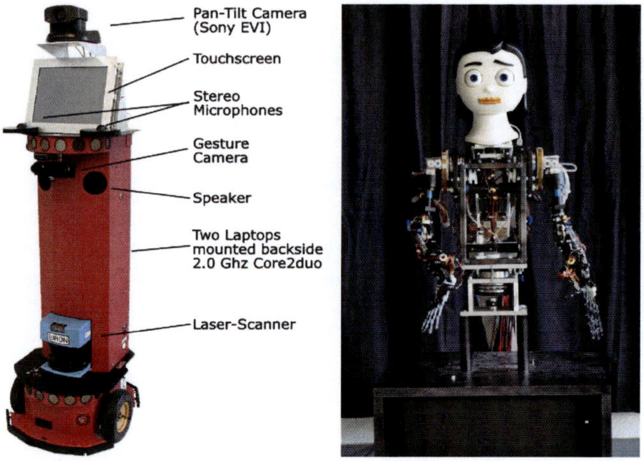

(a) (b)

Fig. 1 The robot demonstrators BIRON (a) and BARTHOC (b). The presented approach has successfully been applied to both of them for advanced HRI.

robot reacts with the same modalities by verbal and gestural output. Taking advantage of a more human-like appearance as BIRON emotional information based on prosody taken from the user's utterances are processed and replied to by adequate facial expressions of BARTHOC [6].

Event Driven Information Exchange

To realize enhanced module communication with ESIA, the concept of information-driven integration is exploited. A module within the system represents either an event source, an event sink or even both of them. Event sources publish their information without a specific receiver, while event sinks subscribe to specific events containing information they need to filter and process [8]. Keeping the focus on the continuous development of a complex system the event-driven information exchange strongly supports the loose coupling of the system components. An event source is not directly connected of its corresponding event sinks and thus it does not matter if a sink is deactivated during runtime. Vice versa if an event source is missing it appears to the sink as if there is currently no new data.

However, though the event-based architectures provide clear advantages for modular systems, they do not directly support persistence. But persistence plays an important role especially for flexible module configuration during runtime. Consider, e.g., an event which is important for the arbitration of an hardware actuator. Since the event source is not aware of its sinks it will not notice if one is currently down, and the event sink will not know about the current arbitration result after a restart. Therefor we combine the event-driven approach with a memory [13]. Besides the

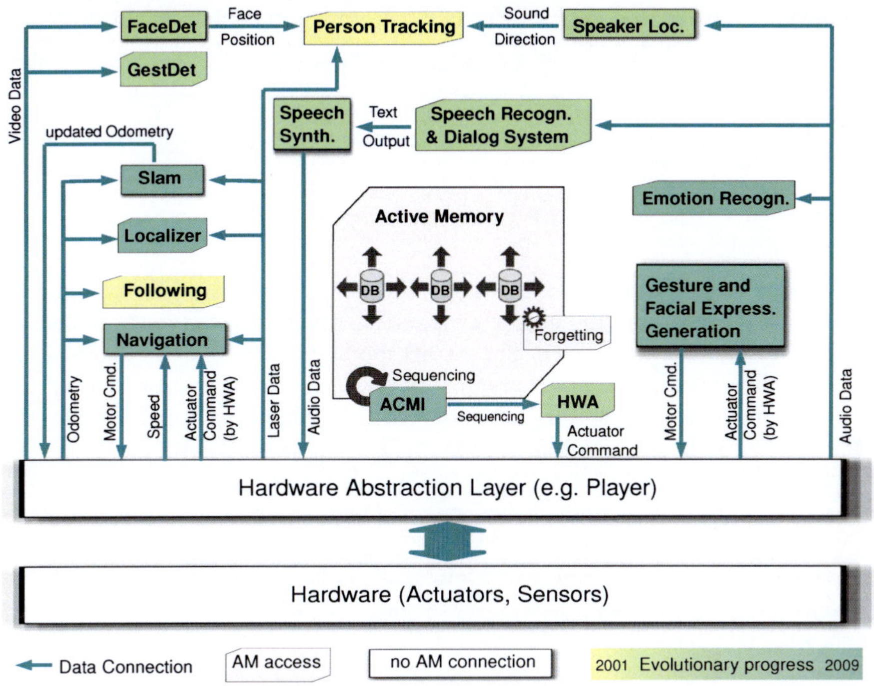

Fig. 2 ESIA for BIRON and BARTHOC: The approach enables a coordinated integration for software which has been developed over the last eight years

event notification, the event information is temporarily stored so it can simply be looked up by a restarted module to get in sync with the remaining system.

An Architecture for HRI

A more concrete description of the implemented architecture with the realized software components for the scenarios and demonstrators is shown in Fig. 2. It is easy to see that both, the reuse of domain unspecific abilities, and the exchange of domain dependent abilities are supported. While raw sensor and actuator information are passed by via direct 1:n connections, processed data is event driven exchanged by the memory, which may contain several XML databases. Using a wide spread information representation such as XML, memory entries are easy addressable by existing tools like XML Path Language (XPath). These tools are also used, to specify the events a module registers to. In principle a registration implements a database trigger, which consists of a database action. If an event takes place associated with an XML document matching the XPath and the according action, the event and the XML document are reported to the registered module. But a robot for enhanced HRI needs a synchronized global system behavior consisting of separate module behaviors to perform complex tasks like learning by interaction. Instead of adding more system knowledge to the individual modules which would have coupled them stronger together, we developed a system keeper module which interfaces with the memory, supervising the information exchange taking place.

4 Active Control Memory Interface - ACMI

One often followed control approach is to establish a direct configuration connection to any component running. This is useful when important information has to be processed under guarantee. The ACMI module supports direct control connection by implementing a generic remote method invocation interface, which has to be supported by the configured module. The interface is used to configure the arbitration of hardware components and provides immediate feedback to ensure the correct configuration before actuator commands are executed. But for a general approach this method contradicts the idea of loose coupling concerning the exchange or addiction of modules during runtime. The general operation method of ACMI is to change the memory content in a way that modules observing the memory react properly to generate the desired system behavior without direct connection to ACMI (see Fig. 3).

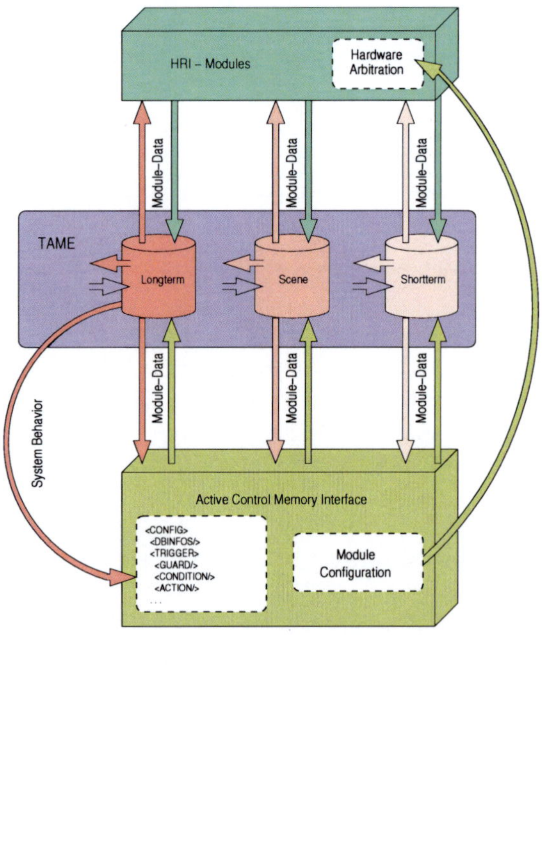

Fig. 3 ACMI operates by manipulating memory content. Exceptionally direct control via RMI is possible. Using the rule editor TAME, the operation of ACMI can be changed immediately during runtime.

Rule Interpretation and Adaptation

ACMI operates by evaluating the current memory content and thus the current system state by a set of rules. An example is given in Fig. 4. It can be compared to rule interpretors of production systems similar to ACT-R [1], but without general control connections to other system components. For the rule interpretation and appliance it reuses the same event based framework as the remaining modules. To enable a fast and easy familiarization of the developers, the amount of different rule types and their actions was reduced to a minimal set. To create full operational readiness five different actions callable by a rule were identified:

copy	copies information from one to another entry
replace	replaces information by information from another entry
remove	removes information in the memory
assert	modifies the current process status of an entry
configure	uses a memory entry as parameter for a RMI of a module

As it can easily be seen from the example a rule is represented the same way, as the information the rules apply to. This does not only keep the number of information representations a developer has to understand small, but additionally allows a kind of reuse of the framework: Instead of separating the rules which implement the system behavior from the factual knowledge, they are stored in the memory as an XML document as an important part of of the system information. Accordingly factual data derived from sensor data and procedural knowledge emerging from rule appliance are treated equally within the memory. This enables the manipulation of rules as easy as any kind of database content even during runtime. Taking into account

```
<CONFIG>
  <ACMI>
    <TRIGGER>
      <GUARD database="Scene" xpath="/SITUATION[@value='object']" exists="no" />
      <GUARD database="Scene" xpath="/SITUATION[@value='localize']" exists="no" />
      <CONDITION database="Scene" xpath="/ROOM[GENERATOR='DLG']/STATUS[@value='initiated']"
                 action="INSERT"/>
      <ACTION name="CONFIGURE">
        <SOURCE database="LongTerm" xpath="/CONFIG/HWC/LOCATING/MOVESRC"/>
      </ACTION>
      <AFFIRM value="accepted"/>
    </TRIGGER>
      ...
    <TRIGGER>
      <GUARD database="Scene" xpath="/SITUATION[@value='object']" exists="no" />
      <GUARD database="Scene" xpath="/SITUATION[@value='localize']" exists="no" />
      <CONDITION database="Scene" xpath="/ROOM/STATUS[@value='accepted']" action="REPLACE"/>
      <ACTION name="REPLACE">
        <SOURCE database="LongTerm" xpath="/CONFIG/SYSTEM/SITUATION[@value='localize']"
                node="/CONFIG/SYSTEM/SITUATION[@value='localize']"/>
        <TARGET database="Scene" xpath="/SITUATION" node="/SITUATION"/>
      </ACTION>
    </TRIGGER>
  </ACMI>
</CONFIG>
```

Fig. 4 Example of ACMI configuration rule and sequencing: When a room description is initially inserted by user dialog and the system is neither in a object detection nor localization task the hardware arbitration is configured to receive a new command from the localization. Subsequently the AFFIRM value is used to modify the STATUS value of the room description, which causes the next rule to set the current system state to localization and prevent other actions to take hardware control before the current task is completed.

that ACMI modifies database content, the described approach enables modification of the system not only during runtime from a human user, but also by the system itself. Thus ESIA provides a powerful basis for learning systems not only on the level of factual knowledge like objects, faces or surroundings, but also on how to behave by modifying a given basic set of rules to a more improved one according to the current operation conditions.

5 Case - Study and Conclusion

To demonstrate the advantage of the presented approach we will briefly summarize the experiences made during two integration processes conducted with partners from alternative universities located at different countries [9, 3]. Two different localization approaches based on omni-directional vision on the one hand and classical laser-based SLAM on the other hand were integrated. As preparatory work the original foreign source was remotely installed and compiled in our lab. Subsequently this integration procedure, generally applicable to ESIA was followed: First, it was determined whether the new module represented an information sink, source, or both and which process loop was used. Second, interfaces for data exchange (according general data models, modules preferably already use a non binary communication) were clarified. Third, rules depending on the outcomes of the first to steps were composed to use ACMI as adapter. Finally, it was checked for conflicts due to e.g. limited resources like actuators and arbitration as well as behavior rules were added to ACMI.

Although during the integration process it turned out that some agreements could not be realized as planned due to technical limitations or simply misunderstandings, the flexibility of the described approach allowed us to continue by simply adjusting or adding rules. As an expected result the demonstrator with its recently extended software modules was successfully tested during user studies concerning the usability of the system. It was robustly running for hours while operated by naive users. When an irritating behavior for the user occurred (in fact the robot prompted some information during interaction which should originally assist the user but in fact confuses several subjects) it was simply modified by changing behavior rules without restarting a single module. This adaptation could actually have been done even during an interaction, although it was not, to preserve experimental conditions for the test person.

The ESIA approach supported this results by providing a short integration cycle powered by a memory allowing detail introspection in data exchange, observing the system behavior by a memory editor and fixing it by changing rules online, and quick restart of single failing modules with the remaining modules continuing unaffected. This enabled us to achieve the described integration together with our partners visiting the lab in three days, both times. The positive experiences and feedback from our partners encourage us to continue in the development of ESIA as evolution is a continuous and never ending process.

Acknowledgements. This work has partially been supported by the German Research Foundation in the Collaborative Research Center (SFB 673) "Alignment in Communication" and the "German Service Robotic Initiative (DESIRE)". Furthermore the authors want to thank all the contributors for providing their time and results. Without the pleasant cooperation the creation of integrated systems would not be possible.

References

1. Anderson, J.R.: Rules of the Mind. Lawrence Erlbaum Associates Inc., Philadelphia (1993)
2. Beuter, B., Spexard, T., Lütkebohle, I., Peltason, J., Kummert, F.: Where is this? - gesture based multimodal interaction with an anthropomorphic robot. In: Proc. of Int. Conf. on Humanoid Robots, Daejeon, Korea (2008)
3. Booij, O., Kröse, B., Peltason, J., Spexard, T.P., Hanheide, M.: Moving from augmented to interactive mapping. In: Proc. of Robotics: Science and Systems Conf., Zurich (2008)
4. Goldin, D., Srinivasa, S., Srikanti, V.: Active databases as information systems. In: Proc. of Int. Database Engineering and Applications Symposium, Washington, DC, pp. 123–130 (2004)
5. Hawes, N., Wyatt, J., Sloman, A.: Exploring design space for an integrated intelligent system. In: Knowledge-Based Systems (2009)
6. Hegel, F., Spexard, T.P., Vogt, T., Horstmann, G., Wrede, B.: Playing a different imitation game: Interaction with an empathic android robot. In: Proc. of Int. Conf. on Humanoid Robots, pp. 56–61 (2006)
7. Ishida, T., Sasaki, T., Nakata, K., Fukuhara, Y.: A meta-level control architecture for production systems. Trans. on Knowl. and Data Eng. 7(1), 44–52 (1995)
8. Lütkebohle, I., Schäfer, J., Wrede, S.: Facilitating re-use by design: A filtering, transformation, and selection architecture for robotic software systems. In: Proc. of Workshop an Software Development in Robotics (2009)
9. Peltason, J., Siepmann, F.H., Spexard, T.P., Wrede, B., Hanheide, M., Topp, E.A.: Mixed-initiative in human augmented mapping. In: Proc. of Int. Conf. on Robotics and Automation (to appear)
10. Qiao, Y., Zhong, K., Wang, H., Li, X.: Developing event-condition-action rules in real-time active database. In: Proc. of Symposium on Applied computing, New York, USA, pp. 511–516 (2007)
11. Santoro, C.: An erlang framework for autonomous mobile robots. In: Proc. of SIGPLAN workshop on ERLANG, New York, USA, pp. 85–92 (2007)
12. Schmidt-Rohr, S.R., Knoop, S., Lösch, M., Dillmann, R.: A probabilistic control architecture for robust autonomy of an anthropomorphic service robot. In: International Conference on Cognitive Systems, Karlsruhe, Germany (2008)
13. Wrede, S.: An information-driven architecture for cognitive systems research. Ph.D. dissertation, Technical Faculty, Bielefeld University (2009)

Direct Control of an Active Tactile Sensor Using Echo State Networks

André Frank Krause, Bettina Bläsing, Volker Dürr, and Thomas Schack

Abstract. Tactile sensors (antennae) play an important role in the animal kingdom. They are also very useful as sensors in robotic scenarios, where vision systems may fail. Active tactile movements increase the sampling performance. Here we directly control movements of the antenna of a simulated hexapod using an echo state network (ESN). ESNs can store multiple motor patterns as attractors in a single network and generate novel patterns by combining and blending already learned patterns using bifurcation inputs.

1 Introduction

Animals and humans are autonomous mobile systems of prime performance, partly due to their highly adaptive locomotor behaviour, partly due to their sensory abilities that allow for rapid, parallel object recognition and scene analysis. In animals, near-range sensing, particularly the active tactile sense is often of great importance:

André Frank Krause · Bettina Bläsing · Thomas Schack
Neurocognition and Action - Biomechanics Research Group,
Bielefeld University, Germany
e-mail: {andre_frank.krause,bettina.blaesing,
thomas.schack}@uni-bielefeld.de

Volker Dürr
Dept. for Biological Cybernetics, University of Bielefeld, Faculty of Biology, PO Box 100131, 33501 Bielefeld, Germany
e-mail: volker.duerr@uni-bielefeld.de

André Frank Krause · Bettina Bläsing · Volker Dürr · Thomas Schack
Cognitive Interaction Technology, Center of Excellence, University of Bielefeld, 33615 Bielefeld, Germany

H. Ritter et al. (Eds.): Human Centered Robot Systems, COSMOS 6, pp. 11–21.
springerlink.com

many insects actively move their antennae (feelers) and use them for orientation, obstacle localisation, pattern recognition and even communication [20]; mammals like cats or rats use active whisker movements to detect and scan objects in the vicinity of the body. Active tactile sensors offer some advantages over vision based sensors [1]: The tactile sense is independent of light conditions, it works at day and night. It is also independent of the surface properties of objects (colour, texture, reflectivity) that may be very difficult for vision, radar or ultrasound based systems. Furthermore, the 3d spatial contact position with an object is immediately available due to the known geometry of the sensor. No computationally expensive stereovision algorithms are required. A drawback of tactile sensors might be lower information density about a scanned object and the danger of mechanical damage to both the sensor and the object. Insect-like tactile sensors have been pioneered by Kaneko and co-workers, who used either vibration signals [26] or bending forces [11], both measured at the base of a flexible beam, to determine contact distance. Recently an actively movable tactile sensor inspired by a stick insect antenna was co-developed by Fraunhofer IFF Magdeburg [16] and the University of Bielefeld [2] with a single acceleration sensor located at the tip of the probe. It was shown to be remarkably sensitive in object material classification. An efficient movement pattern that maximises obstacle detection performance while minimising energy consumption for such an active tactile sensor is a circular movement of the sensor probe [14]. Stick insect antennae perform elliptical exploratory movement patterns relative to the head until they encounter an obstacle. After the first contact, the movement switches to a pattern with smaller amplitude and higher cycle frequency [15].

A straightforward way to learn motor patterns is to store them in the dynamics of recurrent neuronal networks [25]. The network implements a forward model, that predicts the sensor informations for the next time step [28]. In [25] it is argued, that this distributed storage of multiple patterns in a single network gives good generalisation compared to local, modular neural network schemes [3][23]. In [33] it was shown that it is also possible to not only combine already stored motor patterns into new ones, but to establish an implicit functional hierarchy by using leaky integrator neurons with different time constants in a single network. This network can then generate and learn sequences over stored motor patterns and combine them to form new complex behaviours. Tani [25][24][33] uses backpropagation through time (BBTT, [30]), that is computationally complex and rather biologically implausible. Echo State Networks (ESNs [10]) [6] are a new kind of recurrent neuronal networks that are very easy and fast to train compared to classic, gradient based training methods (backpropagation through time [30], real time recurrent learning [32]). Gradient based learning methods suffer from bifurcations that are often encountered during training. Bifurcations abruptly change the dynamic behaviour of a network, rendering gradient information invalid [7]. Additionally it was mathematically shown that it is very difficult to learn long term correlations because of vanishing or exploding gradients [4]. The general idea behind ESNs is to have a large, fixed random reservoir of recurrently and sparsely connected neurons. Only a linear readout layer that taps this reservoir needs to be trained. The reservoir transforms usually low-dimensional, but temporally correlated input signals into a rich feature vector of the reservoir's internal activation dynamics. This is similar to a linear finite impulse response filter or Wiener filter [31], that reads out a tap delay line with a linear combiner. Here, the delay line acts as a preprocessor that constructs a sufficiently large state space from the input time series, such that the temporal dependencies become implicit. Batch training involves collecting the internal states of the reservoir neurons and applying fast linear regression methods to calculate the output layer. More biologically plausible online learning methods for ESNs exist, for example the recursive least squares algorithm [6] or backpropagation decorrelation (BPDC [21]). BPDC was successfully applied in a developmental learning scenario with the humanoid robot ASIMO [18] and for learning an inverse model of the industrial PA-10 robot arm [17]. Here, we use an ESN to implement a forward model that actively moves the tactile sensor / antenna of a simulated hexapod walker.

2 Simulations

A basic, discrete-time, sigmoid-unit ESN was implemented in C++ using the expression template matrix library Eigen2. The state update equations used are:

$$\begin{aligned} \mathbf{y}(n) &= \mathbf{W}^{out}\mathbf{x}(n) \\ \mathbf{x}(n+1) &= \tanh(\mathbf{W}^{res}x(n) + \mathbf{W}^{in}\mathbf{u}(n+1) + \mathbf{W}^{back}\mathbf{y}(n) + \nu(n)) \end{aligned} \tag{1}$$

where $\mathbf{u}$, $\mathbf{x}$ and $\mathbf{y}$ are the activations of the input, reservoir and output neurons, respectively. $\nu(n)$ adds a small amount of uniformly distributed noise to the activation values of the reservoir neurons. This tend to stabilize solutions especially in models using output feedback for cyclic attractor learning [8]. $\mathbf{W}^{in}$, $\mathbf{W}^{res}$, $\mathbf{W}^{out}$ and $\mathbf{W}^{back}$ are the input, reservoir, output and backprojection weight matrices. All matrices are sparse and randomly initialised and stay fixed, except for $\mathbf{W}^{out}$. The weights of the linear output layer are learned using offline batch training. During training, the network is driven with the input and teacher data and internal reservoir activations are collected (state harvesting). The teacher data is forced into the network via the backprojection weights (teacher forcing). After collecting internal states for all training data, The output weights are directly calculated using ridge regression. Ridge regression uses the Wiener-Hopf solution $\mathbf{W}^{out} = \mathbf{R}^{-1}\mathbf{P}$ and adds a regularization term (Tikhonov regularization):

$$\mathbf{W}^{out} = (\mathbf{R} + \alpha^2\mathbf{I})^{-1}\mathbf{P} \tag{2}$$

where α is a small number, $\mathbf{I}$ is the identity matrix, $\mathbf{R} = \mathbf{S}'\mathbf{S}$ is the correlation matrix of the reservoir states and $\mathbf{P} = \mathbf{S}'\mathbf{D}$ is the cross-correlation matrix of the states and the desired outputs. Ridge regression leads to more stable solutions and smaller output weights, compared to ESN training using the Moore-Penrose pseudoinverse.

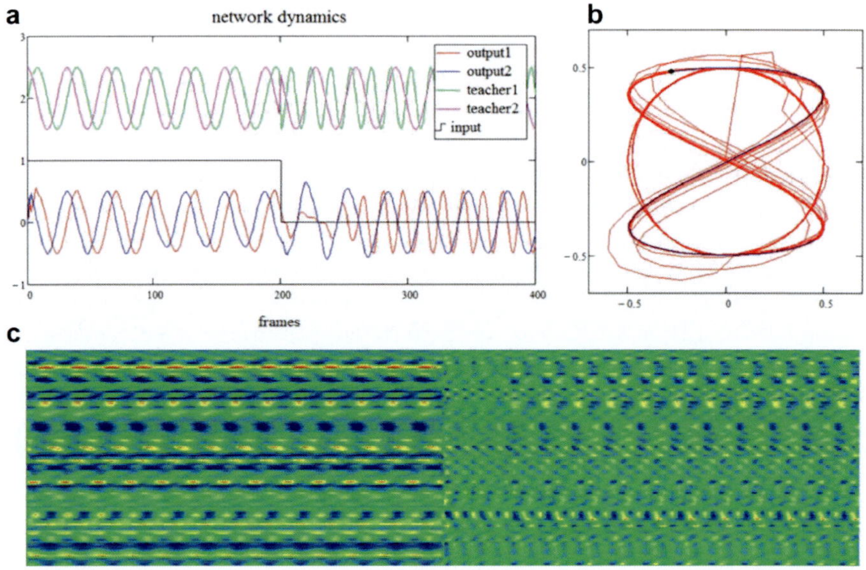

Fig. 1 Dynamic behaviour of an ESN with a single input trained on a circular (input value = 1) and figure-eight pattern (input = 0). a) After 200 time-steps, the input value is switched from one to zero. The network smoothly changes its behaviour to the figure-eight pattern. b) 2d trajectories of the network output. c) Colour coded internal activations of the reservoir neurons. Lines indicate individual neuroids, columns indicate time.

Table 1 ESN structure parameters. 200 reservoir neurons, 2 inputs and 2 outputs used. Direct input to output connections and bias inputs were not used. Sparseness is the fraction of connections with non-zero weights. Synaptic weights were randomly initialised in the range [-strength strength]. $\alpha = 0.01$, $\nu = 0.001$.

from	to	Sparseness	Strength
Input	Reservoir	1.0	0.6
Reservoir	Reservoir	0.1	0.4
Output	Reservoir	1.0	0.8

Dynamic Behaviour. The input-, reservoir- and backprojection weight matrices were sparsely initialised with uniformly distributed random values. See table 1 for the network parameters used. In a first simulation, a simple ESN with one input, two outputs, no bias inputs and no input-to-output connections was trained on the circle (input value = 1) and figure-eight pattern (input value = 0) (see table 3). Fig. 1 shows the dynamic behaviour of this network. Abruptly switching the input from one to zero smoothly moves the network from the circular pattern attractor state to the figure-eight pattern. In Fig. 2 the input parameter space is explored within the range -1.0 to 1.2. Interestingly, an input value of -1 evokes the same circular pattern as for an input value of one. This symmetric effect is mentioned in [7]. Increasing the input value further causes gradual morphing from a circular to an elliptical and later to the figure-eight pattern. Increasing the input above 1 or below -1 drives the network into a fixpoint attractor state.

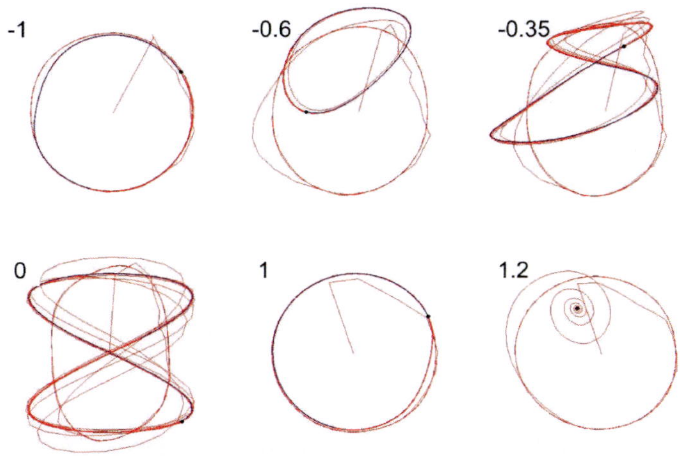

Fig. 2 Shifting the dynamics of the network by gradually changing the input value from -1 to 1.2. Because of symmetry effects [7], the circular pattern reappears with input value -1. Increasing the input to the network causes a slow morphing between the two learned patterns, allowing to generate new patterns that were not explicitly trained. The network keeps stable with no chaotic regions until it converges to a fixpoint at input value 1.2.

Training Success. In a second simulation, the training success and the transition behaviour between two patterns after switching the input values was analysed. Instead of a single input, this time 2 binary encoded inputs were used. The first pattern was learned with an input vector $\begin{pmatrix} 1 & 0 \end{pmatrix}$ and the second with $\begin{pmatrix} 0 & 1 \end{pmatrix}$. This avoids symmetry effects as shown in simulation 1 and reduces the training error. ESN parameters used are shown in table 1. The training error was defined as the smallest Euclidean Distance between the training pattern and a time-shifted version of the network output (± 200 time-steps). The selected time-shift corresponded to the largest cross-correlation coefficient. The error of a single network was averaged over n=50 runs. Input patterns were presented in random order and all network activations were randomly intialized in a range of ± 0.4 before each run. N=500 networks were trained, resulting in a median error of 0.029 (0.6% deviation relative to the circle pattern radius of 0.5). 60% of the trained networks had an error below 1%.

Transition Smoothness. After switching between patterns, the network might become unstable and show chaotic behavior. Relating to the Minimum Jerk Theory

Table 2 ESN structure parameters. 300 reservoir neurons, 4 inputs and 2 outputs were used in the multiple pattern storage task. $\alpha = 0.001$, $\nu = 0.001$.

from	to	Sparseness	Strength
Input	Reservoir	1.0	0.8
Reservoir	Reservoir	0.1	0.2
Output	Reservoir	1.0	1.2

[5], the smoothness of the movement after the switch can be quantified as a function of jerk, which is the time derivative of acceleration. The jerk was calculated for 100 timesteps after the switch. The mean jerk for 500 networks averaging over 50 runs for each net was 0.024 with a standard deviation of 0.003. This is just slightly larger than the averaged jerk of both training patterns (0.0196). The transition behaviour was sufficiently stable for all 500 networks, the maximum jerk found was 0.038. For comparison, mean jerk of purely random, untrained networks was 41.6 with a SD of 40.713.

Learning Capacity. A larger ESN was trained with four different patterns, see table 3. ESN parameters used are shown in table 2. Fig. 3 shows that it is possible to store multiple motor patterns distributed in a single network. Nonetheless it requires manual parameter fine-tuning to get stable attractors that are close to the training data.

Table 3 Training patterns used for the ESN experiments

Pattern	Parameters
Circle	$0.5\left(\sin(0.2n)\ \cos(0.2n)\right)$
Eight	$0.5\left(\sin(0.4n)\ \sin(0.2n)\right)$
Rectangle	$0.5\left(\tanh(2\sin(0.2n))\ \tanh(2\cos(0.2n))\right)$
Star	$0.2\left(\text{atanh}(0.98\sin(0.2n))\ \text{atanh}(0.98\cos(0.2n))\right)$

Motor Control. Stick insects continuously move their antennae during walking using a wide, exploratory movement pattern. If the antennae detect obstacles, the antennal movements immediately change to a sampling pattern [15]. This switching behaviour was modeled using an ESN and a simulated hexapod walker with antennae. The simulation was implemented in C++ using the Open Dynamics Engine (ODE). The joints of the antenna were steered using a p-controller and constraint-based angular velocity control (hinge joint angular motor in ODE). Due to the dynamic nature of the system and the p-controller, actual joint angles always lag some frames behind the desired values and have a slightly smaller amplitude. The network thus has to learn to predict new motor commands based on the proprioreceptive input from the antennal joints. In a first step, training data was created by sinusoidal

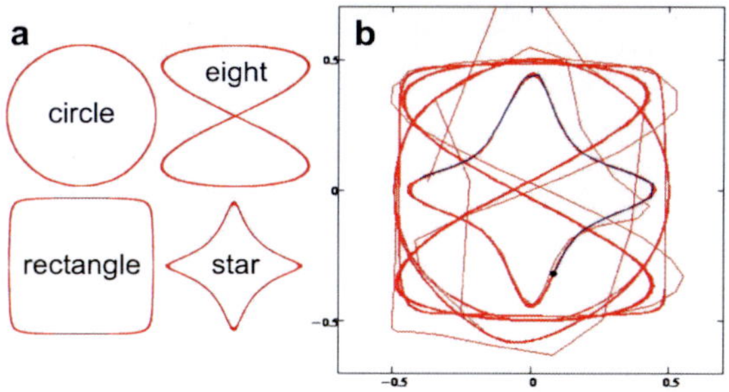

Fig. 3 An ESN with 4 binary inputs, two outputs and 300 reservoir neurons was trained to store multiple patterns. a) shows the 4 patterns used as training data (for parameters see table 3. b) shows the network dynamics. in the generation phase, the network was started with random network activations and simulated for 4000 time steps. the input value changed every 1000 time steps.

modulation of the antennal joints and simultaneously recording actual joint angles in the simulation, see Fig. 4. An ESN was then trained on the collected data and put into the loop (identical network parameters as in the initial experiment, see table 1). If the left antenna encountered contacts with obstacles, the input values to the ESN were switched, causing a transition from a wide, exploratory movement pattern to the figure-eight pattern. After some decay time, the inputs were switched back. Fig. 5 shows a behaviour sequence of an hexapod walker with an ESN-controlled left antenna. Obstacle contact causes a pre-programmed obstacle avoidance reflex by turning the insect away from the object.

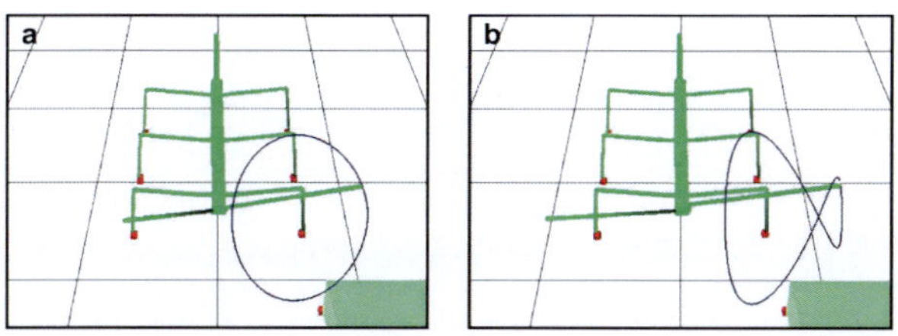

Fig. 4 An ESN was trained to generate a circular and a figure eight pattern with the tip of the left antenna. Inputs to the net are the current joint angles and the pattern selection inputs; outputs are the target joint angles for the p-controller steering the joints of the active tactile sensor.

3 Discussion and Outlook

First basic experiments with ESNs have shown that they can be used for direct motor control of simple articulated limbs. The ESN implements a forward model that predicts sensory and motor commands for the next time step. It is also possible to

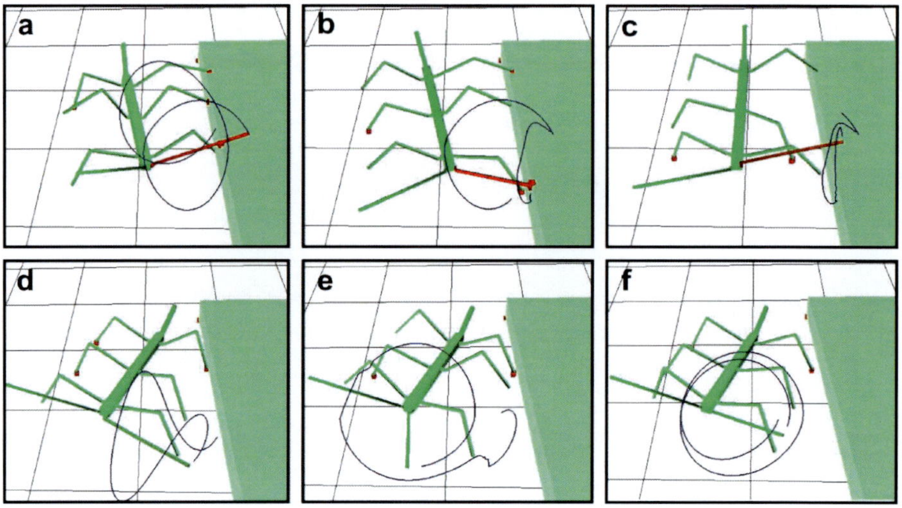

Fig. 5 Sequence of images showing antennal behaviour before, during and after contacts with an obstacle. From left to right: a) first contact of the left antenna with the obstacle. The antenna still performs a circular, exploratory movement. b) the contact information causes a switch in behaviour from the circular to the figure-eight pattern. c) Second and third contact: the hexapod walker starts to turn away from the obstacle. d) The figure-eight pattern continues for a while until the contact information decays. e-f) After that, the behaviour switches back to the exploratory circular movement. A video can be downloaded at: http://www.andre-krause.net/publications/hcrs09_s1.avi

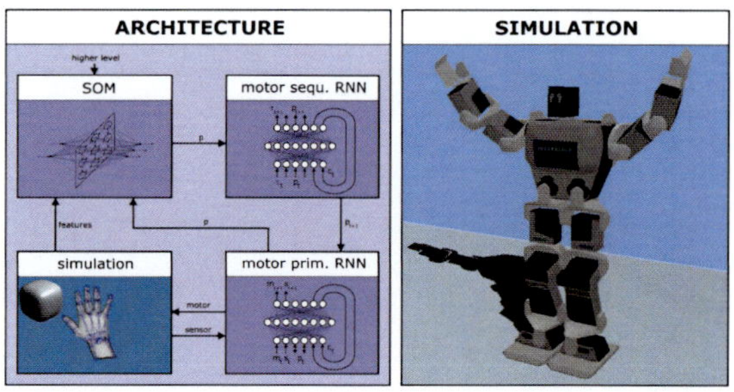

Fig. 6 A proposed architecture for learning and generation of complex movements. Hierarchically coupled ESNs are controlled using a hierarchical self organizing map, that implements basic action concepts. Image: ©Webots [29].

generate new, not explicitly trained patterns by shifting the network dynamics through additional bifurcation inputs. This was already demonstrated by [25] via parametric bias inputs for a variant of Elman type networks. If exploited properly, this dynamic feature of ESN networks makes it possible to generate and interpolate numerous motor patterns from a few, well chosen basic motor patterns. ESNs can also store multiple motor patterns in a single network, although it is important to fine-tune all network parameters to succeed. Pretraining of the reservoir using Intrinsic Plasticity [22] can help to make the training process more robust. ESN parameters could also be automatically fine-tuned in a reinforcement scenario using new, very fast and powerful black box optimisation algorithms [12] [27]. ESNs seem to be suitable for a planned hierarchical architecture for learning and control of long, complex movement sequences, as illustrated in Fig. 6. ESNs scale well to a high number of training patterns and motor outputs [9]. A more complex simulation - for example of a humanoid robot - might reveal possible limitations of direct, attractor-based motor pattern learning. The future goal is to couple two or more ESNs hierarchically or even embed an implicit hierarchy into a single network using neurons with random time constants similar to [33]. On top of that, a hierarchical self-organizing map (HSOM) can implement cognitive structures of basic action concepts [19] [13] and provide input and reference values to the ESNs. The HSOM can integrate perceptual features of the environment, proprioceptive sensory data of the robot body and higher level commands (intention, affordance) to select a proper motor program. Cluster structures learned in the HSOM might then be compared to cognitive structures that can be measured in human long term memory using a psychological method called SDA-M [19].

References

1. Dürr, V., Krause, A.: The stick insect antenna as a biological paragon for an actively moved tactile probe for obstacle detection. In: Berns, K., Dillmann, R. (eds.) Climbing and walking robots - from biology to industrial applications, Proceeding of Fourth International Conference Climbing and Walking Robots (CLAWAR 2001), pp. 87–96. Professional Engineering Publishing, Bury St. Edmunds (2001)
2. Dürr, V., Krause, A.F., Neitzel, M., Lange, O., Reimann, B.: Bionic tactile sensor for near-range search, localisation and material classification. In: Berns, K., Luksch, T. (eds.) Autonome Mobile Systeme 2007. Fachgespräch Kaiserslautern, vol. 20, pp. 240–246. Springer, Heidelberg (2007)
3. Haruno, M., Wolpert, D.M., Kawato, M.: Mosaic model for sensorimotor learning and control. Neural Computation 13(10), 2201–2220 (2001)
4. Hochreiter, S., Bengio, Y.: Gradient flow in recurrent nets: the difficulty of learning long-term dependencies. In: Kremer, S.C., Kolen, J.F. (eds.) A Field Guide to Dynamical Recurrent Neural Networks. IEEE Press, Los Alamitos (2001)
5. Hogan, N.: An organizing principle for a class of voluntary movements. Journal of Neuroscience 4, 2745–2754 (1984)
6. Jaeger, H.: Adaptive nonlinear system identification with echo state networks. In: Becker, S., Thrun, S., Obermayer, K. (eds.) Advances in Neural Information Processing Systems, vol. 15, pp. 593–600. MIT Press, Cambridge (2002)
7. Jaeger, H.: Tutorial on training recurrent neural networks, covering bppt, rtrl, ekf and the "‘echo state network"‘ approach. Tech. Rep. GMD Report 159, German National Research Center for Information Technology (2002)
8. Jaeger, H., Lukosevicius, M., Popovici, D., Siewert, U.: Optimization and applications of echo state networks with leaky integrator neurons. Neural Networks 20(3), 335–352 (2007)

9. Jäger, H.: Generating exponentially many periodic attractors with linearly growing echo state networks. Technical report 3, IUB (2006)
10. Jäger, H., Haas, H.: Harnessing nonlinearity: Predicting chaotic systems and saving energy in wireless communication. Science 304, 78–80 (2004)
11. Kaneko, M., Kanayma, N., Tsuji, T.: Active antenna for contact sensing. IEEE Transactions on Robotics and Automation 14, 278–291 (1998)
12. Kramer, O.: Fast blackbox optimization: Iterated local search and the strategy of powell. In: The 2009 International Conference on Genetic and Evolutionary Methods, GEM 2009 (in press, 2009)
13. Krause, A.F., Bläsing, B., Schack, T.: Modellierung kognitiver Strukturen mit hierarchischen selbstorganisierenden Karten. In: Pfeffer, I., Alfermann, D. (eds.) 41. Jahrestagung der Arbeitsgemeinschaft für Sportpsychologie (asp), vol. 188, p. 91. Czwalina Verlag Hamburg (2009)
14. Krause, A.F., Dürr, V.: Tactile efficiency of insect antennae with two hinge joints. Biological Cybernetics 91, 168–181 (2004)
15. Krause, A.F., Schütz, C., Dürr, V.: Active tactile sampling patterns during insect walking and climbing. In: Proc. Göttingen Neurobiol. Conf., vol. 31 (2007)
16. Lange, O., Reimann, B.: Vorrichtung und Verfahren zur Erfassung von Hindernissen. German Patent 102005005230 (2005)
17. Reinhart, R.F., Steil, J.J.: Attractor-based computation with reservoirs for online learning of inverse kinematics. In: European Symposium on Artificial Neural Networks (ESANN) – Advances in Computational Intelligence and Learning (2009)
18. Rolf, M., Steil, J.J., Gienger, M.: Efficient exploration and learning of whole body kinematics. In: IEEE 8th International Conference on Development and Learning (2009)
19. Schack, T., Mechsner, F.: Representation of motor skills in human long-term memory. Neuroscience Letters 391, 77–81 (2006)
20. Staudacher, E., Gebhardt, M.J., Dürr, V.: Antennal movements and mechanoreception: neurobiology of active tactile sensors. Adv. Insect. Physiol. 32, 49–205 (2005)
21. Steil, J.J.: Backpropagation - decorrelation: online recurrent learning with o(n) complexity. In: Proceedings of the 2004 IEEE International Joint Conference on Neural Networks, vol. 2, pp. 843–848 (2004)
22. Steil, J.J.: Online reservoir adaptation by intrinsic plasticity for backpropagation-decorrelation and echo state learning. Neural Networks 20(3), 353–364 (2007)
23. Tani, J., Nolfi, S.: Learning to perceive the world as articulated: an approach for hierarchical learning in sensory-motor systems. Neural Networks 12, 1131–1141 (1999)
24. Tani, J.: On the interactions between top-down anticipation and bottom-up regression. Frontiers in Neurorobotics 1, 2 (2007)
25. Tani, J., Itob, M., Sugitaa, Y.: Self-organization of distributedly represented multiple behavior schemata in a mirror system: reviews of robot experiments using RNNPB. Neural Networks 17, 1273–1289 (2004)
26. Ueno, N., Svinin, M., Kaneko, M.: Dynamic contact sensing by flexible beam. IEEE/ASME Transactions on Mechatronics 3, 254–264 (1998)
27. Vrugt, J., Robinson, B., Hyman, J.: Self-adaptive multimethod search for global optimization in real-parameter spaces. IEEE Transactions on Evolutionary Computation 13(2), 243–259 (2008)
28. Webb, B.: Neural mechanisms for prediction: do insects have forward models? Trends in Neuroscience 27, 278–282 (2004)
29. Webots: Commercial Mobile Robot Simulation Software, `http://www.cyberbotics.com`

30. Werbos, P.: Backpropagation through time: what it does and how to do it. Proceedings of the IEEE 78(10), 1550–1560 (1990)
31. Wiener, N.: Extrapolation, interpolation, and smoothing of stationary time series with engineering applications. Cambridge, Technology Press of Massachusetts Institute of Technology and New York, Wiley (1949)
32. Williams, R.J., Zipser, D.: A learning algorithm for continually running fully recurrent neural networks. Neural Computation 1(2), 270–280 (1989), http://www.mitpressjournals.org/doi/abs/10.1162/neco.1989.1.2%.270
33. Yamashita, Y., Tani, J.: Emergence of functional hierarchy in a multiple timescale neural network model: A humanoid robot experiment. PLoS Computational Biology 4(11) (2008)

Dynamic Potential Fields for Dexterous Tactile Exploration

Alexander Bierbaum, Tamim Asfour, and Rüdiger Dillmann

Abstract. Haptic exploration of unknown objects is of great importance for acquiring multimodal object representations, which enable a humanoid robot to autonomously execute grasping and manipulation tasks. In this paper we present our ongoing work on tactile object exploration with an anthropomorphic five-finger robot hand. In particular we present a method for guiding the hand along the surface of an unknown object to acquire a 3D object representation from tactile contact data. The proposed method is based on the dynamic potential fields which have originally been suggested in the context of mobile robot navigation. In addition we give first results on how to extract grasp affordances of unknown objects and how to perform object recognition based on the acquired 3D point sets.

1 Introduction

Humans make use of different types of haptic exploratory procedures for perceiving physical object properties such as weight, size, rigidity, texture and shape [12]. For executing subsequent tasks on previously unknown objects such as grasping and also for non-ambiguous object identification the shape property is of the utmost importance. In robotics this information is usually obtained by means of computer vision where known weaknesses such as changing lightning conditions and reflections seriously limit the scope of application. For robots and especially for humanoid robots, tactile perception is supplemental to the shape information given by visual perception and may directly be exploited to augment and stabilize a spatial representation of real world objects. In the following we will give a short overview on the state of the art in the field of robot tactile exploration and related approaches.

Alexander Bierbaum · Tamim Asfour · Rüdiger Dillmann
University of Karlsruhe (TH), Institute for Anthropomatics, Humanoids and Intelligence Systems Laboratories
e-mail: {bierbaum,asfour,dillmann}@ira.uka.de

H. Ritter et al. (Eds.): Human Centered Robot Systems, COSMOS 6, pp. 23–31.
springerlink.com

Different strategies for creating polyhedral object models from single finger tactile exploration have been presented with simulation results in [19] and [5]. Experimental shape recovery results from a surface tracking strategy for a single robot finger have been presented in [6]. A different approach concentrates on the detection of local surface features [15] from tactile sensing. In [13] a method for reconstructing shape and motion of an unknown convex object using three sensing fingers is presented. In this approach friction properties must be known in advance and the surface is required to be smooth, i.e. must have no corners or edges. Further, multiple simultaneous sensor contacts points are required resulting in additional geometric constraints for the setup.

In the works mentioned above the exploration process is based on dynamic interaction between the finger and object, in which a sensing finger tracks the contour of a surface. Other approaches are based on a static exploration scheme in which the object gets enclosed by the fingers and the shape is estimated from the robot finger configuration. In [14], [9] and [20] the finger joint angle values acquired during enclosure are fed to an appropriately trained SOM-type neural network which classifies the objects according to their shape. Although this approach gives good results in terms of shape classification, it is naturally limited in resolution and therefore does not provide sufficient information for general object identification as with dynamic tactile exploration.

In this work we will present the current state and components of our system for acquiring a 3D shape model of an unknown object using multi-fingered tactile exploration based on dynamic potential fields. In addition we give first results on how to extract grasp affordances of unknown objects and how to perform object recognition based on the acquired 3D point sets.

2 Dynamic Potential Fields for Exploration

We have transferred the idea of potential field based exploration to tactile exploration for surface recovery using an anthropomorphic robot hand. Potential field techniques have a long history in robot motion planning [11]. Here, the manipulator follows the streamlines of a field where the target position is modelled by an attractive potential and obstacles are modelled as repulsive potentials. By assigning regions of interest to attractive sources and already known space to repulsive sources this scheme may also be exploited for spatial exploration purposes with mobile robots [18]. The notion of dynamic potential fields evolves as the regions of interest and therefore the field configuration changes over time due to the progress in exploration. Yet, this method has not been reported for application in multifingered tactile exploration. For this purpose we have defined a set of *Robot Control Points* (RCPs) at the robot hand to which we apply velocity vectors calculated from the local field gradient

$$\mathbf{v} = -k_v \nabla \Phi(x).$$

The potential $\Phi(x)$ is calculated from superposition of all sources. We use harmonic potential functions to avoid the formation of local minima in which the imaginary force exerted on an RCP is zero. Further, we deploy a dedicated escape strategy to resolve structural minima, which naturally evoke from the multiple end-effector problem given by the fingers of the hand. The velocity vectors applied to the RCPs are computed in the cartesian coordinate frame therefore an inverse kinematic scheme is required to calculate joint angles for the robot hand and arm. In our case we have chosen Virtual Model Control (VMC) [17] to solve for the joint angles, as it links the potential field approach to inverse kinematics in an intuitive way.

Initially we have evaluated our approach in a detailed physical simulation using the model of our humanoid robot hand [8]. During exploration the contact location and estimated contact normals are acquired from the robot hands tactile sensor system and stored as a oriented 3D point set. We have modelled tactile sensors in the simulation environment which determine contact location. The contact normals are estimated from the sensor orientation to reflect the fact that current sensor technology can not measure contact normals reliably. The object representation may be used for further applications such as grasping and object recognition as we will describe in the following sections.

3 Tactile Exploration

Fig. 1 gives an overview on our tactile exploration module. An initial version of this method has been presented in [3]. As prerequisite the system requires a rough initial estimate about the objects position, orientation and dimension. In simulation we introduce the information to the system, while this information will be provided by a stereo camera system in the real application. From this information an initial potential field containing only attractive sources is constructed in a uniform grid which covers the exploration space in which the object is situated.

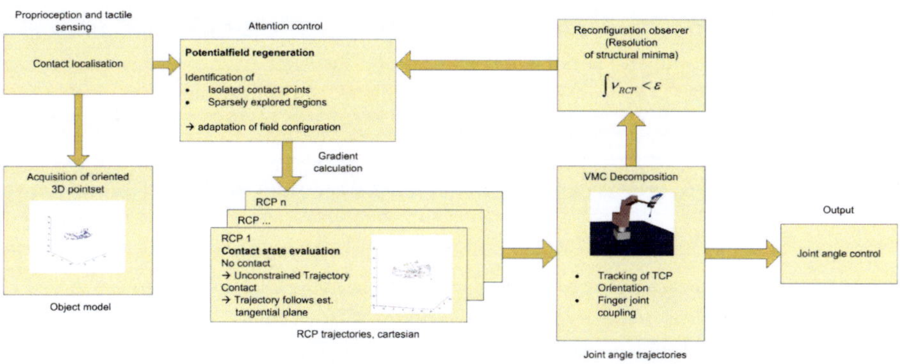

Fig. 1 Tactile exploration scheme based on dynamic potential field

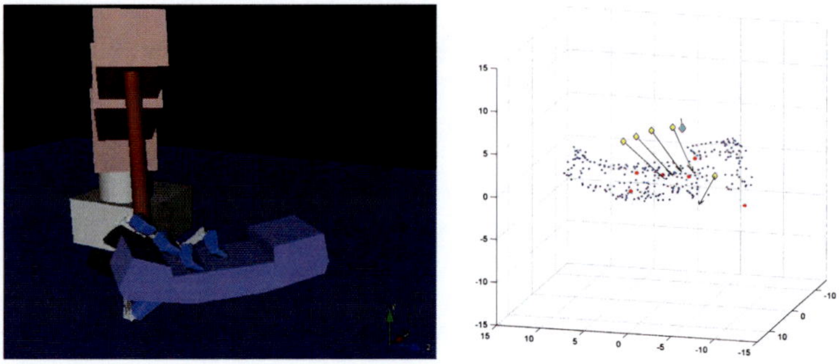

Fig. 2 Tactile exploration of a phone receiver (left) and acquired 3D point set (right)

During exploration it is required to fixate the object as contact points are acquired in world reference frame. The trajectories for the RCPs become continuously calculated from the field gradient, while contact point locations and normals are sensed and stored as oriented 3D point set. The normal vectors are estimated from finger sensor orientations. The RCP trajectories are constrained depending on the contact state of the sensor associated with each RCP, which aims to produce tangential motion during contact.

The potential field is updated from the tactile sensor information as follows. If no contact is found in the circumference of an attractive source it becomes deleted. If a contact is detected a repelling source is inserted at the corresponding location in the grid.

The robot system is likely to reach structural minima during potential field motion. We therefore introduced a reconfiguration observer which detects when the TCP velocity and the mean velocity of all RCPs fall below predefined minimum velocity values. This situation leads to a so called *small reconfiguration* which is performed by temporarily inverting the attractive sources to repulsive sources and thus forcing the robot into a new configuration which allows to explore previously unexplored goal regions. As this method is not guaranteed to be free of limit cycles we further perform a *large reconfiguration* if subsequent small reconfigurations remain ineffective, i.e. the robot does not escape the structural minimum. During a large configuration the robot is moved to its initial configuration.

Our approach to extract grasp affordances relies on identifying suitable opposing and parallel faces for grasping. Therefore, we needed to improve the original tactile exploration process to explore the object surface with preferably homogeneous density and prevent sparsely explored regions. The faces become extracted after applying a triangulation algorithm upon the acquired 3D point set. Triangulation naturally generates large polygons in regions with low contact point count. We use this property in our improved exploration scheme to introduce new attractive sources and guide the exploration process to fill contact information gaps. Within fixed time step intervals we execute a full triangulation of the point cloud and rank the calculated faces by their size of area. In our modification we add an attractive source each at the centers of the ten largest faces. This leads to preferred exploration of sparsely explored regions, i.e. regions that need further exploration, and consequently lead to a more reliable estimate for the objects surface. As further improvement we apply a similar scheme to isolated contact points, i.e. contacts which have no further contact points in their immediate neighborhood, by surrounding these points with eight cubically arranged attractive charges. This leads to the effect that once an isolated contact is added, the according RCP now explores its neighborhood instead of being repelled to a more distant unexplored region.

4 Extraction of Grasp Affordances

As an exemplary application for our exploration procedure we have implemented a subset of the automatic robot grasp planner proposed in [16] in order to compute possible grasps based on the acquired oriented 3D point set, we call *grasp affordances*. A grasp affordance contains a pair of object features which refer to grasping points of a promising grasp candidate using a parallel grasp. We preferred to investigate this geometrical planning approach in contrast to grasp planning algorithms using force closure criteria, e.g. [7], due to its robustness when planning with incomplete geometric object models as they arise from the described exploration scheme. In our case we only consider planar face pairings from the given 3D point set as features for grasping, which we extract from the contact normal vector information using a region growing algorithm. Initially every possible face pairing is considered as a potential symbolic grasp. All candidates are submitted to a geometric filter pipeline which eliminates impossible grasps from this set. The individual filter j returns a value of $f_o, j = 0$ when disqualifying and a value $f_o, j > 0$ for accepting a pairing. For accepted pairings the individual filter outputs are summed to a score for each symbolic grasp, where the filter pairing with the highest score is the most promising candidate for execution.

The filter pipeline comprises the following stages in order of their application.

- *Parallelism:* This filter tests the two faces for parallelism and exports a measure indicating the angle between the two faces.
- *Minimum Face Size:* This filter compares the two faces to minimum and maximum thresholds. Selection of these values depends on the dimensions of the robot hand and fingers.
- *Mutual Visibility:* This filter determines the size of overlapping area when the two faces are projected into the so called grasping plane, which resides in parallel in the middle between the faces.
- *Face Distance:* This filter tests the distance of the two faces which must match the spreading capability of the robot hand. Therefore, this filter is also parameterized by the dimensions of the robot hand.

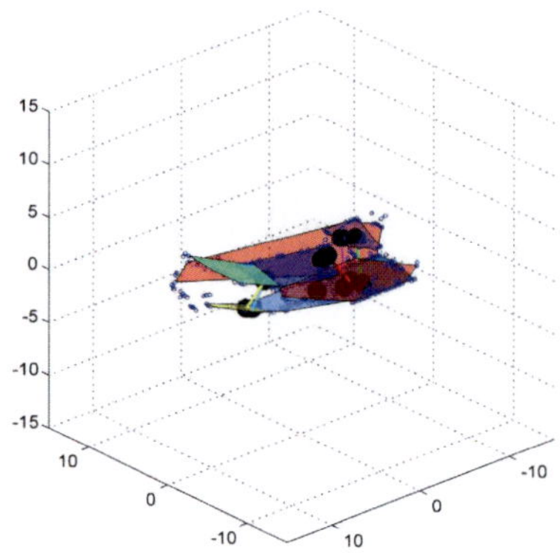

Fig. 3 Extracted grasp affordances for the telephone receiver

Fig. 3 shows symbolic grasps found for the receiver from Fig. 2. Face pairings are indicated by faces of the same color, the black spots mark the centers of the overlapping region of opposing faces in respect to the grasping plane. These points will later become the finger tip target locations during grasp execution.

5 Future Concepts for Object Recognition

The oriented 3D point set acquired from tactile exploration is inherently sparse and of irregular density which makes shape matching a difficult task. In a first approach we have investigated a superquadric fitting technique which allows to estimate a super quadric function from tactile contacts in a robust manner [2]. Fig. 4 (left)

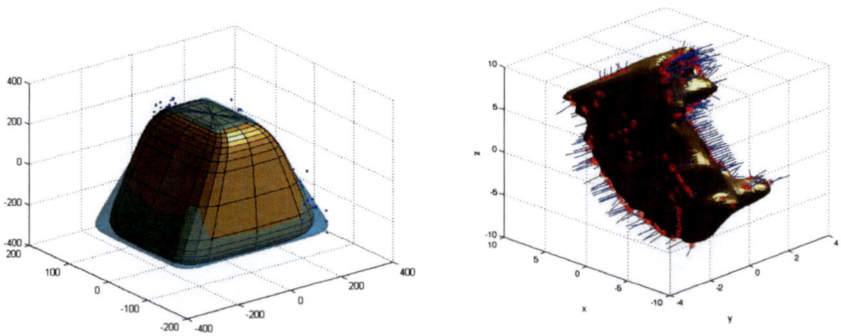

Fig. 4 Superquadric reconstructed from a tactile point set (left). A surface reconstructed using 3D Fourier transform (right).

shows a superquadric recovered from tactile exploration data using a hybrid approach where a genetic algorithm is used to identify the global minimum region and a least-squares-method converges to an optimum solution. Yet, this method is limited to representing and recognizing shapes only from a set of major geometric primitives such as spheres, cylinders, boxes or pyramids. For representing more complex shapes, different shape descriptors which may also become applied to partial models have been investigated in the research fields of computer vision and 3D similarity search [4]. The methods reported are mainly designed for large 3d data sets with uniform sampling density. Therefore, we have focused on investigating suitable point set processing methods which may interpolate the tactile contact data in order to compute robust shape descriptors. Fig. 4 (right) shows an oriented point set from tactile exploration which has been interpolated by using an algorithm for reconstruction of solid models [10]. From uniform density point sets stable shape descriptors may be computed using methods developed in the context of computer vision. Promising candidates for distinct shape descriptors here are geometric hash tables and spectra from spherical harmonic transforms. Both provide means for translational and rotational invariance, which is essential in object recognition from exploration data in human environments.

6 Discussion

In this paper we presented an overview on our system for tactile exploration. Our approach is based on dynamic potential fields for motion guidance of the fingers of a humanoid hand along the contours of an unknown object. We added a potential field based reconfiguration strategy to eliminate structural minima which may arise from limitations in configuration space. During the exploration process oriented point sets from tactile contact information are acquired in terms of a 3D object model. Further, we presented concepts and preliminary results for applying the geometric object model to extract grasp affordances from the data. The grasp affordances comprise grasping points of promising configurations which may be executed by a robot using parallel-grasps. For object recognition we have outlined our approach which relies on transforming the sparse and non-uniform point set from tactile exploration to a model representation appropriate for 3D shape recognition methods known from computer vision.

We believe that the underlying 3D object representation of our concept is a major advantage as it provides a common basis for multimodal sensor fusion with a stereo vision system and other 3D sensors. As finger motion control during exploration is directly influenced from the current model state via the potential field, this approach becomes a promising starting point for developing visuo-haptic exploration strategies.

Currently we extend our work in several ways. In a next step we will transfer the developed tactile exploration scheme to our robot system Armar-III [1] which is equipped with five-finger hands and evaluate the concept in a real world scenario.

Further, we are developing and implementing a motion controller which is capable to execute and verify the grasp affordances extracted from exploration. For object recognition we will continue to investigate suitable shape descriptors and evaluate them with simulated and real world data from tactile exploration.

References

1. Asfour, T., Regenstein, K., Azad, P., Schroder, J., Bierbaum, A., Vahrenkamp, N., Dillmann, R.: Armar-iii: An integrated humanoid platform for sensory-motor control. In: 6th IEEE-RAS International Conference on Humanoid Robots, pp. 169–175 (2006), doi:10.1109/ICHR.2006.321380
2. Bierbaum, A., Gubarev, I., Dillmann, R.: Robust shape recovery for sparse contact location and normal data from haptic exploration. In: IEEE/RSJ 2008 International Conference on Intelligent Robots and Systems, Nice, France, pp. 3200–3205 (2008), doi:10.1109/IROS.2008.4650982
3. Bierbaum, A., Rambow, M., Asfour, T., Dillmann, R.: A potential field approach to dexterous tactile exploration. In: International Conference on Humanoid Robots 2008, Daejeon, Korea (2008)
4. Bustos, B., Keim, D.A., Saupe, D., Schreck, T., Vranić, D.V.: Feature-based similarity search in 3d object databases. ACM Comput. Surv. 37(4), 345–387 (2005), `http://doi.acm.org/10.1145/1118890.1118893`
5. Caselli, S., Magnanini, C., Zanichelli, F., Caraffi, E.: Efficient exploration and recognition of convex objects based on haptic perception. In: Proceedings of the 1996 IEEE International Conference on Robotics and Automation, vol. 4, pp. 3508–3513 (1996), doi:10.1109/ROBOT.1996.509247
6. Chen, N., Rink, R., Zhang, H.: Local object shape from tactile sensing. In: Proceedings of the 1996 IEEE International Conference on Robotics and Automation, vol. 4, pp. 3496–3501 (1996), doi:10.1109/ROBOT.1996.509245
7. Ferrari, C., Canny, J.: Planning optimal grasps. In: Proceedings of the 1992 IEEE International Conference on Robotics and Automation, vol. 3, pp. 2290–2295 (1992), doi:10.1109/ROBOT.1992.219918
8. Gaiser, I., Schulz, S., Kargov, A., Klosek, H., Bierbaum, A., Pylatiuk, C., Oberle, R., Werner, T., Asfour, T., Bretthauer, G., Dillmann, R.: A new anthropomorphic robotic hand. In: IEEE RAS International Conference on Humanoid Robots (2008)
9. Johnsson, M., Balkenius, C.: Experiments with proprioception in a self-organizing system for haptic perception. In: Wilson, M.S., Labrosse, F., Nehmzow, U., Melhuish, C., Witkowski, M. (eds.) Towards Autonomous Robotic Systems, pp. 239–245. University of Wales, Aberystwyth (2007)
10. Kazhdan, M.: Reconstruction of solid models from oriented point sets. In: SGP 2005: Proceedings of the third Eurographics symposium on Geometry processing, vol. 73, Eurographics Association, Aire-la-Ville, Switzerland, Switzerland (2005)
11. Khatib, O.: Real-time obstacle avoidance for manipulators and mobile robots. The International Journal of Robotics Research 5(1), 90–98 (1986)
12. Lederman, S.J., Klatzky, R.L.: Hand movements: A window into haptic object recognition. Cognitive Psychology 19(3), 342–368 (1987)

13. Moll, M., Erdmann, M.A.: Reconstructing the Shape and Motion of Unknown Objects with Active Tactile Sensors, ch. 17. Springer Tracts in Advanced Robotics, pp. 293–310. Springer, Heidelberg (2003)
14. Natale, L., Metta, G., Sandini, G.: Learning haptic representation of objects. In: International Conference on Intelligent Manipulation and Grasping, Genoa, Italy (2004)
15. Okamura, A., Cutkosky, M.: Haptic exploration of fine surface features. In: Proceedings of the 1999 IEEE International Conference on Robotics and Automation, vol. 4, pp. 2930–2936 (1999), doi:10.1109/ROBOT.1999.774042
16. Pertin-Troccaz, J.: Geometric reasoning for grasping: a computational point of view. In: Ravani, B. (ed.) CAD Based Programming for Sensory Robots. NATO ASI Series, vol. 50. Springer, Heidelberg (1988)
17. Pratt, J., Torres, A., Dilworth, P., Pratt, G.: Virtual actuator control. In: Proc. IEEE/RSJ International Conference on Intelligent Robots and Systems, IROS 1996, vol. 3, pp. 1219–1226 (1996), doi:10.1109/IROS.1996.568974
18. Silva, P.E., Engel, P.M., Trevisan, M., Idiart, M.A.P.: Exploration method using harmonic functions. Robotics and Autonomous Systems 40(1), 25–42 (2002)
19. Roberts, K.: Robot active touch exploration: constraints and strategies. In: Proceedings of the 1990 IEEE International Conference on Robotics and Automation, vol. 2, pp. 980–985 (1990), doi:10.1109/ROBOT.1990.126118
20. Takamuku, S., Fukuda, A., Hosoda, K.: Repetitive grasping with anthropomorphic skin-covered hand enables robust haptic recognition. In: Proc. IEEE/RSJ International Conference on Intelligent Robots and Systems IROS, pp. 3212–3217 (2008), doi:10.1109/IROS.2008.4651175

Unlimited Workspace - Coupling a Mobile Haptic Interface with a Mobile Teleoperator

Thomas Schauß, Ulrich Unterhinninghofen, and Martin Buss

Abstract. A teleoperation system extends cognitive and manipulatory skills of a human to a remote site. Hereby, the dimensions of the haptic interfaces and telemanipulators typically limit the achievable workspace. In some application scenarios, however, an unrestricted workspace is desirable. Therefore, we develop and discuss two different coupling schemes for a wide-area teleoperation system. The schemes are implemented on a complex system composed of two haptic interfaces and two telemanipulator arms mounted on mobile bases. Aside from haptic feedback also visual and acoustic feedback are presented in an appropriate way. Experiments show the benefits and drawbacks of the different schemes, and conclusions on appropriate coupling schemes for different circumstances are given.

1 Introduction

Although autonomous robots have come a long way in the past years, they are not yet capable of performing complex tasks in unknown, unstructured environments. Using a haptic teleoperation system, an operator can control a robot located at a remote site. Thus, the benefits of robots and the cognitive skills of humans can be merged.

Since its invention in the 1940s, teleoperation has been a field of active research. While most hardware and control architectures are focused on low-power, small-workspace applications, e.g. telesurgery and microassembly [2], teleoperation also has many important applications in human scale environments, e.g. disaster recovery and remote maintenance. In these applications, the used workspace matches or even exceeds the workspace of the human arm.

Thomas Schauß · Ulrich Unterhinninghofen · Martin Buss
Institute of Automatic Control Engineering, Technische Universität München, D-80290 München, Germany
e-mail: schauss@tum.de, ulrich.unterhinninghofen@tum.de, m.buss@ieee.org

H. Ritter et al. (Eds.): Human Centered Robot Systems, COSMOS 6, pp. 33–41.
springerlink.com

Among the various techniques for haptic interaction in large-scale environments [3], only mobile haptic interfaces and force exoskeletons can provide a truly unlimited workspace. However, the latter are cumbersome and fatiguing due to their heavy weight. Design and control guidelines for mobile haptic interfaces have been developed in [4] and extended in [1]. We have shown how to optimally position a bimanual mobile haptic interface in order to maximize manipulability and workspace in [8]. A full teleoperation system for a single arm based on a mobile haptic interface and mobile teleoperator has been examined in [5].

In this work, a haptic teleoperation system for applications in extensive environments is presented. The system comprises a mobile haptic interface and a mobile teleoperator, which are both designed to enable bimanual manipulations in six degrees of freedom. The mobile haptic interface consists of two large-scale haptic interfaces which are mounted on an omnidirectional mobile base. The design of the mobile teleoperator is similar such that two anthropomorphic robot arms are mounted on top of a mobile base.

The focus of this paper lies on the interconnection scheme between mobile haptic interface and mobile teleoperator. Two different approaches are presented and compared regarding their performance for different applications. In Sec. 2, the interconnection schemes are described. Sec. 3 offers on overview of the employed hardware components. The design and the results of the performance experiments are presented in Sec. 4. The paper concludes with Sec. 5, where a summary and an outlook on future work are given.

2 Proposed Methods

In Fig. 1, the components of the teleoperation system are depicted along with the associated coordinate systems. The position and orientation of the mobile haptic interface, i.e. the master, ${}^0T_M$, the mobile teleoperator, i.e. the slave, ${}^0T_S$, and the human operator ${}^0T_H$ are expressed in a common world coordinate system Σ_0. The end-effector poses of the right and the left haptic interface are denoted ${}^0T_R$ and ${}^0T_L$, and the respective interaction forces[1] are denoted ${}^0F_R$ and ${}^0F_L$. Analogously, the end-effector poses of the robot-arms on slave side are ${}^0T_R^*$ and ${}^0T_L^*$, while the forces are ${}^0F_R^*$ and ${}^0F_L^*$.

A typical teleoperation system aims at synchronizing master and slave poses as well as master and slave forces. Assuming that no transformation, e.g. scaling, occurs between master and slave coordinates, this aim can be expressed for right and left end-effectors by the four equality relations:

$$ {}^0T_R = {}^0T_R^* \qquad {}^0T_L = {}^0T_L^* \qquad {}^0F_R = {}^0F_R^* \qquad {}^0F_L = {}^0F_L^* \tag{1} $$

[1] Throughout this paper, forces represent generalized forces, i.e. translational forces and rotational torques.

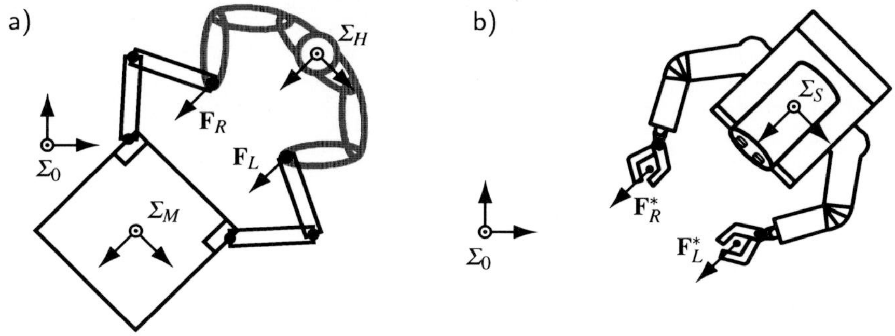

Fig. 1 Definition of coordinate systems of a) master side and b) slave side

As haptic interfaces and robot arms are mounted on mobile bases, the end-effector poses and interaction forces are expressed relative to the coordinate system of the mobile base on master side Σ_M and slave side Σ_S, respectively. Thus, the signals on master side become ${}^M T_R$ and ${}^M T_L$ as well as ${}^M F_R$ and ${}^M F_L$. Likewise, the signals on slave side are locally expressed by ${}^S T_R^*$ and ${}^S T_L^*$ as well as ${}^S F_R^*$ and ${}^S F_L^*$.

Both mobile devices have redundant degrees of freedom. When the poses of left and right end-effectors are given, this does not yield a unique solution for the pose of the mobile base. Any solution which produces the desired end-effector position in world-coordinates is equally valid:

$$ {}^0T_M\, {}^M T_R = {}^0 T_R = {}^0 T_R^* = {}^0 T_S\, {}^S T_R^* \qquad {}^0T_M\, {}^M T_L = {}^0 T_L = {}^0 T_L^* = {}^0 T_S\, {}^S T_L^*. \quad (2)$$

There are different criteria which can be used to find the desired pose of the mobile bases ${}^0T_{M,d}$ and ${}^0T_{S,d}$. In general, the optimization criteria must take the different dynamic capabilities of the mobile bases and the haptic interfaces or robot arms, respectively, into account. As the mobile bases carry the whole system weight, their maximum velocity and acceleration are comparatively low. On the other hand, the workspace of the mobile base is theoretically unrestricted, whereas the workspace of the haptic interfaces and the robot arms is limited. Consequently, large-scale motions with low dynamics must be assigned to the mobile bases, whereas fast, small-scale motions must be performed by the haptic interfaces or robot-arms.

For the mobile haptic interface, two optimization methods for finding the optimal pose of the mobile base are discussed in [8]. Below, similar methods for the mobile teleoperator are presented.

2.1 Coupling Schemes

In this section, two methods for the mobile teleoperator which determine the optimal pose of its mobile base are discussed. The operating principles are depicted in Fig. 2.

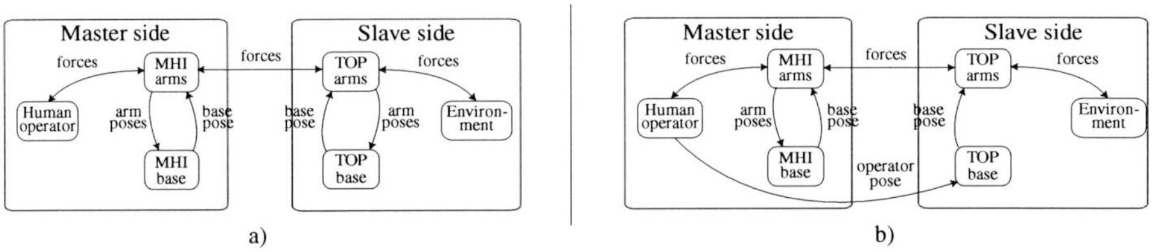

Fig. 2 Coupling schemes for mobile haptic interface and mobile teleoperator. In scheme a) only the end-effectors are coupled and the position of the mobile base on slave side is optimized, whereas in scheme b) the mobile base is positioned depending on the tracked position of the operator.

One method uses an optimization scheme in order to determine the pose of the mobile base as a function of the end-effector poses (Fig. 2a)). The second method relies on tracking the body of the human operator and replicating its position by the teleoperator (Fig. 2b)).

Coupling of End-Effectors Only

The first proposed coupling scheme (Fig. 2a)) consists of optimizing the pose of the base of the mobile teleoperator (${}^0T_S$) with respect to the end-effector poses (${}^0T_R^*, {}^0T_L^*$). Different optimization criteria can be considered. The base can e.g. be positioned, so that the two telemanipulator arms have a maximum distance from their respective workspace boundaries or so that the velocity manipulability is maximized. Here, the maximum distance of the end-effectors from the center of the respective workspace is minimized, thereby approximately optimizing the workspace of the telemanipulator arms.[2]

Using this scheme, the pose of the mobile teleoperator (${}^0T_S$) is independent of the pose of the human operator (${}^0T_H$). A disadvantage of this scheme is that locomotion of the operator does not directly result in movement of the mobile base on teleoperator side, and movement of the mobile base on teleoperator side can occur without locomotion of the operator. Therefore, the relative pose of the end-effectors on local side (${}^HT_R^*, {}^HT_L^*$) and remote side (${}^ST_R^*, {}^ST_L^*$) is not necessarily equal, also not in steady state. Thus, a negative effect on task performance and feeling of immersion can be expected. An advantage of this scheme is that the pose of the operator (${}^0T_H$) does not have to be tracked, making an expensive wide-area tracking system unnecessary.

[2] This approximation only takes into account the translational DOF of the telemanipulator arm. The optimization problem for 6 DOF in Cartesian space or 7 DOF in joint space is non-convex and thus not solvable in real-time.

Coupling of Body and End-Effectors

An alternative coupling scheme is depicted in Fig. 2b). Hereby, the mobile teleoperator (${}^0T_S$) is controlled in such a way that it tracks the pose of the operator (${}^0T_H$).

Locomotion of the operator then directly results in movement of the mobile teleoperator. Using this scheme, the point of view and movement of the camera is expected to more closely follow the operators locomotion. In steady state, the pose of the mobile teleoperator (${}^0T_S$) will be identical to the pose of the human operator (${}^0T_H$). Then, also the relative pose of the end-effectors on local side (${}^HT_R^*, {}^HT_L^*$) and remote side (${}^ST_R^*, {}^ST_L^*$) becomes identical. This scheme is expected to provide superior task performance and a higher degree of immersion. However, reaching the workspace limits or a configuration with a low manipulability could occur more easily, as the mobile platform on teleoperator side is not controlled to optimize these criteria. Furthermore, tracking of the human operator (${}^0T_H$) is necessary to determine the control input for the mobile teleoperator.

2.2 Control Structure

In Fig. 3, the interconnection of the involved subsystems is depicted. The teleoperation system is realized as position-based admittance control architecture. The forces on master side (${}^MF_R, {}^MF_L$) and on slave side (${}^SF_R^*, {}^SF_L^*$) are transformed into the world coordinate system by means of the poses of the respective mobile base (${}^0T_M, {}^0T_S$). The transformed forces are summed and fed into two independent admittance equations which implement the desired mass, damping, and optionally stiffness. The outputs of the admittance equations are a desired pose for right and

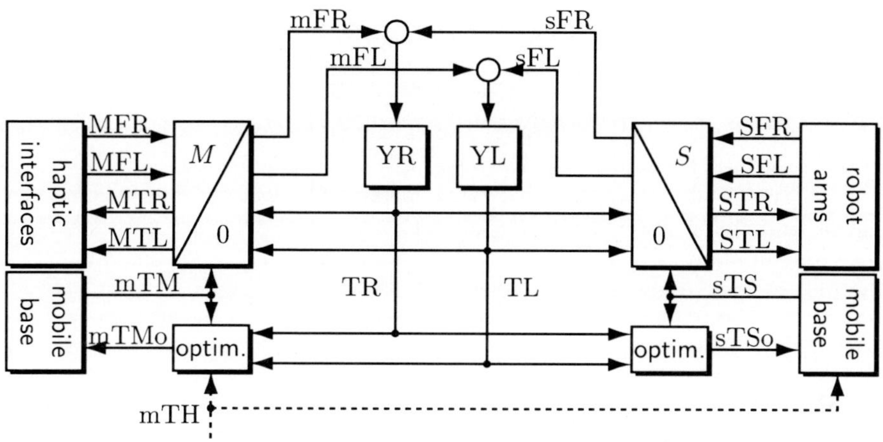

Fig. 3 Overall control structure of the wide-are teleoperation system with position based admittance control. The dashed line indicates the optional tracking data of the operator pose.

left end-effector in world-fixed coordinates ${}^{0}T_{R,d}, {}^{0}T_{L,d}$. These poses are used in two ways. On the one hand, they are transformed back into the local coordinate systems of mobile haptic interface and mobile teleoperator, where they form the control inputs to the haptic interfaces and robot arms, respectively. On the other hand, they are used for determining the optimal pose of the mobile base. Depending on the optimization strategy, additional inputs such as the body pose of the human operator ${}^{0}T_{H}$ are integrated.

3 Setup

Two identical omnidirectional non-holonomic mobile bases are used in this work. Each mobile base is composed of four independently driven and steered wheels mounted on a rectangular frame. The bases are capable of carrying a payload of up to 200 kg. The underlying controller allows independent translational and rotational movement. For more details on hardware and control of the mobile bases, see [4].

On master side, two ViSHaRD7 haptic interfaces are mounted on the mobile base. Each haptic interface has one translational and six rotational joints, resulting in seven actuated DOF. A half-cylinder with a radius and height of approx. 0.6 m and therefore most of the workspace of a human arm is covered. Peak forces of up to 150 N can be output at the end-effectors making it possible to convincingly display stiff remote environments. The kinematic structure is especially designed to decouple translational and rotational DOF. This is beneficial for the optimization of the manipulability of the haptic interfaces as presented in [8]. For more details on mechanical design and control of the ViSHaRD7, see [6].

Two anthropomorphic seven DOF telemanipulator arms are mounted on the mobile base of the slave side. As for the haptic interfaces, each arm is scaled to approximately cover the workspace of a human arm, and a payload of 6 kg can be held in any configuration. For details on design and control of the teleoperator arms, see [7].

In addition, a stereo camera and two microphones are mounted on a pan-roll-tilt base on the mobile teleoperator. The video and audio stream is presented to the operator over a Head Mounted Display (HMD). Thus, a 3D-view and stereo sound of the remote site can be presented to the operator. On operator side, the commercial acoustic tracking system IS-900 by InterSense Inc. is used, which allows measuring the pose of multiple objects in six DOF. Specifically, the position and orientation of mobile base, operator body, and operator head are determined.

4 Results

Two experiments were performed to evaluate the coupling schemes given in Sec. 2. As tracking of the operator position is necessary for the scheme in which the mobile

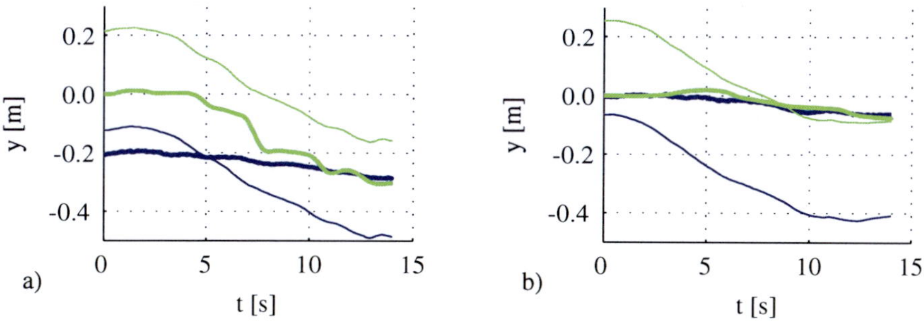

Fig. 4 Experimental results for manipulation while standing still for a) Coupling of the end-effectors only and b) Coupling of body and end-effectors. The bold dark/blue line depicts the position of the operator body while the bold light/green line depicts the base position of the mobile teleoperator. The thin blue and green line are the position of the right and left end-effector respectively. Results are given in the common world coordinate system. ${}^{0}T_S(t=0)$ is chosen to be the identity matrix.

base position is coupled to the operator position, a more sophisticated optimization scheme is used for the mobile haptic interface. This takes the operator position and his workspace boundaries into account, see [8].

The first experiment consists of a manipulation task while standing still. The operator picks up an object to his left (positive y-direction), moves it to the right, and places it down. Hereby he does not move his body more than necessary. Fig. 4 shows the recorded data for the experiment. As can be seen, the body of the operator only moves slightly in both experiments. In Fig. 4a) the case is depicted where only the end-effector positions are coupled. The mobile base is controlled to maintain a symmetric position with respect to both end-effectors[3]. In contrast, Fig. 4b) shows the method where operator body and mobile base are directly coupled. Only little deviation of the base position is observed in this case. This results in a more natural relative end-effector position. However, the right arm is near the workspace boundaries at the end of the experiment.

In the second experiment, the operator moves forwards a given distance and then backwards again to his starting position. While moving forwards, the operator holds his hands close to his body. While moving backwards, he extends his arms. In Fig. 5a) results are illustrated for the case where only end-effector positions are coupled. Again, the movement of the mobile base does not directly dependent on the movement of the operator body. This has the advantage that the end-effectors are far from the workspace boundaries as long as the operator does not move too fast. However, as can be seen well at the turning point, even large position changes of the operator do not result in base movement if the hands are kept still. This results in an unnatural feeling of locomotion as proprioception and visual cues are

[3] The high velocity between second 7 and 8 is caused by the non-holonomic base, which must first adjust the steering angle of the wheels before it can accelerate appropriately.

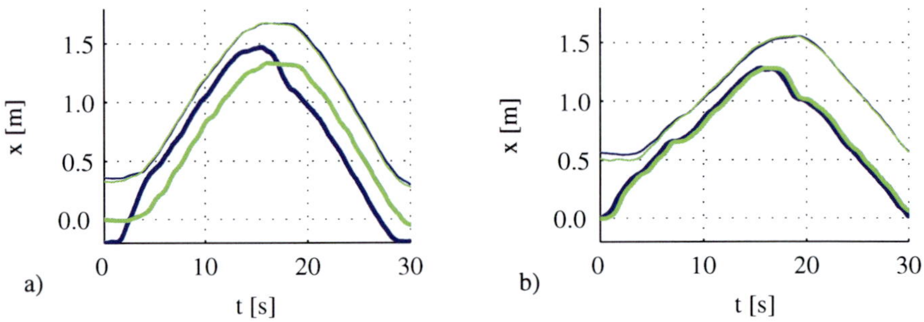

Fig. 5 Experimental results for walking for a) Coupling of the end-effectors only and b) Coupling of body and end-effectors. The bold dark/blue line depicts the position of the operator body while the bold light/green line depicts the base position of the mobile teleoperator. The thin blue and green line are the position of the right and left end-effector respectively. Results are given in the common world coordinate system. ${}^{0}T_{S}(t = 0)$ is chosen to be the identity matrix.

inconsistent. Complementary results are obtained by the second coupling scheme where the movement of operator and mobile base are closely coupled, see Fig. 5b).

5 Conclusion

The concept of a mobile haptic interface and a mobile teleoperator for teleoperation in extensive environments was revisited. Two schemes for coupling the two mobile devices were presented, where the differences lie in the positioning of the mobile bases. In the first mode, which can be implemented without expensive tracking systems, both mobile bases are controlled to a position, which maximizes the distance to the workspace limits of haptic interfaces and robot arms, respectively. In the second coupling mode, the position of the human body relative to his hands is used to derive the base position of the mobile teleoperator. Furthermore, the base position of the mobile haptic interface is optimized for fast manipulation tasks.

In summary, both schemes enable spatially unlimited teleoperation, but it was shown that additional tracking can enhance the operator experience. In contrast to other methods previously presented, these schemes do not sacrifice low operator fatigue for an unlimited workspace and high output capabilities.

Some open questions still remain that could be addressed in the future. It could e.g. be beneficial to combine the two coupling schemes, i.e. optimizing both relative body pose and workspace of the teleoperator. Furthermore, a general solution to avoid workspace limits while maintaining position tracking between haptic interfaces and teleoperator arms has not been developed yet for position-based admittance control schemes as the one presented here.

Acknowledgements. This work is supported in part by the German Research Foundation (DFG) within the collaborative research center SFB453 "High-Fidelity Telepresence and Teleaction" and in part by the European Commission under Contract FP6 034002 ROBOT@CWE.

References

1. Formaglio, A., Prattichizzo, D., Barbagli, F., Giannitrapani, A.: Dynamic performance of mobile haptic interfaces 24(3), 559–575 (2008)
2. Hokayem, F.P., Spong, W.M.: Bilateral teleoperation: An historical survey. Automatica 42(12), 2035–2057 (2006)
3. Künzler, U., Runde, C.: Kinesthetic haptics integration into large-scale virtual environments. In: Proceedings of the First Joint Eurohaptics Conference, and Symposium on Haptic Interfaces for Virtual Environment and Teleoperator Systems. World Haptics 2005 (2005)
4. Nitzsche, N., Hanebeck, U., Schmidt, G.: Design issues of mobile haptic interfaces. Journal of Robotic Systems 20(9), 549–556 (2003)
5. Nitzsche, N., Schmidt, G.: A mobile haptic interface mastering a mobile teleoperator. In: Proc. IEEE/RSJ International Conference on Intelligent Robots and Systems (IROS 2004), vol. 4, pp. 3912–3917 (2004)
6. Peer, A., Komoguchi, Y., Buss, M.: Towards a mobile haptic interface for bimanual manipulations. In: Proceedings of the IEEE/RSJ International Conference on Intelligent Robots and Systems, IROS (2007)
7. Stanczyk, B.: Developement and Control of an Anthropomorphic Teleoperator. Ph.D. thesis, Technische Universität München, Lehrstuhl für Steuerungs- und Regelungstechnik (2006)
8. Unterhinninghofen, U., Schauß, T., Buss, M.: Control of a mobile haptic interface. In: Proc. IEEE International Conference on Robotics and Automation ICRA 2008, pp. 2085–2090 (2008)

An Architecture for Real-Time Control in Multi-robot Systems

Daniel Althoff, Omiros Kourakos, Martin Lawitzky, Alexander Mörtl, Matthias Rambow, Florian Rohrmüller, Dražen Brščić, Dirk Wollherr, Sandra Hirche, and Martin Buss

Abstract. This paper presents a novel robotic architecture that is suitable for modular distributed multi-robot systems. The architecture is based on an interface supporting real-time inter-process communication, which allows simple and highly efficient data exchange between the modules. It allows monitoring of the internal system state and easy logging, thus facilitating the module development. The extension to distributed systems is provided through a communication middleware, which enables fast and transparent exchange of data through the network, although without real-time guarantees. The advantages and disadvantages of the architecture are rated using an existing framework for evaluation of robot architectures.

1 Introduction

Software complexity of the emerging generation of versatile robotic systems increases alongside with their capabilities. Cooperative action of multiple robots, each controlled by numerous software modules, requires seamless message and data exchange internally among the modules, among the robots as well as with distributed sensor systems.

We consider a scenario of multiple robots operating in a populated environment and interacting with humans. Multiple sensors such as tracking systems, on-board laser range finders and force/position sensors permanently generate new data. Data-processing modules such as localization and the generation of a 3-D world representation retrieve sensor data and update the world model. This in turn is utilized by

Daniel Althoff · Omiros Kourakos · Martin Lawitzky · Alexander Mörtl · Matthias Rambow · Florian Rohrmüller · Dražen Brščić · Dirk Wollherr · Sandra Hirche · Martin Buss
Institute of Automatic Control Engineering (LSR), Technische Universität München, D-80290 München, Germany
e-mail: {da,omiros.kourakos,ml,moertl,rambow,rohrmueller, drazen,dw,hirche,mb}@tum.de

H. Ritter et al. (Eds.): Human Centered Robot Systems, COSMOS 6, pp. 43–52.
springerlink.com

high-level reasoning and planning algorithms in order to issue appropriate plans and commands. Finally, low-level processes control actuators in order to execute these commands.

To cope with the challenges arising from this kind of systems sophisticated software frameworks that provide interfaces between the modules have been developed. Popular examples such as *Player* [2] or *ROS* [7] are user-friendly frameworks incorporating a large number of open-source modules which enable a rapid implementation of robotic applications. Even though these software frameworks offer mature basic services in various respects, so far they cannot fully satisfy all requirements of highly modularized and distributed next-generation cognitive robotic systems like:

- support of feedback control under real-time constraints,
- seamless data acquisition and sharing among multiple modules in one robot,
- inter-connectivity and bandwidth for efficient distributed operation,
- elaborate structure to provide maintainability and scalability, and
- user support tools that convey the system state in an intuitive manner and facilitate debugging and data logging.

In this paper we present a software architecture suited for research on interaction and cooperation in complex multi-robot scenarios. The architecture focuses on distributed robotic systems and provides support from real-time execution in low-level control up to efficient high-level information exchange.

This paper is organized as follows. Section 2 presents the proposed architecture and gives a qualitative evaluation that illustrates its strengths and weaknesses. In section 3 a brief application example is given. Finally, section 4 concludes the paper.

2 The *ARCADE* Framework

ARCADE stands for **A**rchitecture for **R**eal-time **C**ontrol and **A**utonomous **D**istributed **E**xecution. It is a data-driven architecture built up on top of a real-time capable interface for inter-process communication and the IceStorm publisher-subscriber extension of the ICE RPC framework [5]. The *ARCADE* framework is illustrated in Fig. 1 and explained in the following.

2.1 System Description

*Real-time database (RTDB):*The real-time database *KogMo-RTDB* [4] – originally developed for autonomous cars – is the central part of the framework. This real-time database is not a database in its traditional meaning, however it implements a number of features of hierarchical databases. The RTDB provides real-time guarantees for management and exchange of data structures, defined as data objects in the RTDB terminology. It handles all local inter-process communication conveniently,

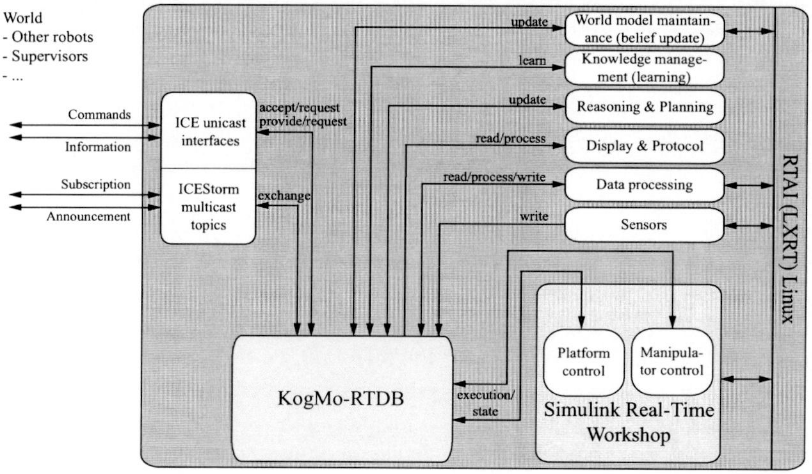

Fig. 1 Overview of the *ARCADE* framework

allows record/replay of states, buffers data and acts as a powerful central hub for all data and command transactions. A further core functionality of the RTDB is the maintenance of hierarchical data objects. Data objects can be easily inserted, updated, deleted and attached as child objects. Searching and reading can be executed in a blocking or non-blocking fashion, depending on the specific needs. In addition, circular buffers of configurable size are maintained to hold histories of data objects. The RTDB offers very high convenience and ease of use for managing complex and large amounts of data while giving real-time assurances. Performance measurements for the database back end throughput and jitter are given in [3], for example are the average/worst case times in a strongly busy system without real-time configuration $23\mu s/181681\mu s$ for write operations and $17\mu s/10721\mu s$ for read operations respectively. With real-time configuration the average times are similar while the worst case times are strongly reduced to $134\mu s$ (write) and $62\mu s$ (read) [3].

Accessing the RTDB: The RTDB can be accessed using the available *ARCADE* interfaces. An interface is a C++ class implementing a set of methods that operate on a data object. Additionally, the *ARCADE* interfaces automatically inherit the RTDB interface methods (read, write, search etc.).

Any process that uses one or more *ARCADE* interfaces is defined as a module. For illustration we describe one producer and one consumer module that use the same interface. These modules could be a driver for a laser range finder and a line extraction algorithm respectively. In this case the interface comprises the data object, which is an array of floating point numbers holding the range measurements, and methods to set (setData) and to get (getData) the data object.

Both modules first have to create a connection to the RTDB. The producer has to instantiate an *ARCADE* interface and insert it in the RTDB with a descriptive name. This name can be used by other modules to search and subscribe to this data. The setData method is used to update the local copy of the data object which subsequently is committed using the write method. The RTDB itself imposes no constraints on the update rate (read/write) of the module which solely depends on the refresh rate of the laser range finder.

The consumer has to instantiate the same interface and associate it with the existing object by searching the RTDB with the descriptive name as explained above. The local copy is updated using the read method and the getData method to access the data object and process it. Both modules can be implemented either as best-effort (i.e. non real-time) processes or real-time tasks using RTAI/LXRT [1].

In addition, the *ARCADE* framework provides library blocks for Simulink to access the RTDB. This enables rapid-prototypical control design in Simulink together with Real-Time Workshop [9] using the RTAI target.

Inter-RTDB communication: In order to transfer data between RTDBs the ZeroC/ICE middleware [5] is integrated into the *ARCADE* framework. ICE provides an operating system independent interface and supports the most common object-oriented programming languages. Furthermore it allows type safe and clear interface definitions. ICE servers give access to any RTDB within the entire system for reading data and setting command objects.

For information exchange among the robots, the IceStorm multicast extension of ICE is used. IceStorm is a publisher/subscriber mechanism for unidirectional information exchange. Any module is able to create and publish topics via IceStorm. Agents on the network can subscribe to topics and receive automatic updates on content changes. In our wireless network setup, bandwidth efficiency is a crucial factor. For the distribution of high-frequency sensor data we extended the IceStorm multicast to allow the specification of a desired update rate alongside with topic subscriptions. The maximum desired update rate for each topic determines the rate of multicast frames sent through the network. Hence, an efficient synchronization of robots in order to maintain a global up-to-date world model is possible.

In the *ARCADE* framework inter-RTDB data exchange can be implemented with corresponding ICE interfaces. However, the abstraction of this procedure remains future work.

Command exchange via the RTDB: The RTDB was originally designed to distribute data streams rather than commands. Straightforward updating of a command object in the database is not possible in case command reception has to be guaranteed. The database command object could be updated more frequently than it is read by the receiver. Due to the finite length of the object history maintained by the RTDB successful transmission cannot be guaranteed and as a consequence commands can be lost. In this respect, to enable the command exchange via the database a three-way handshake – referred to as "postbox system" – has been integrated based on the functionalities provided by the RTDB. It is illustrated in Fig. 2.

A postbox object (P) is created by the receiver side (RX) and then regularly checked for new message objects. A sender process (TX) can then insert a message object, which is attached to the postbox. During its next update cycle the receiver side reads all new messages, which in turn get an acknowledgement object appended. Since only the process which created a specific object obtains write access to it, the sender process regularly checks all its previously sent objects and deletes them as soon as a corresponding acknowledgement object is detected. Consequently this three-way handshake obviates memory leaks and ensures reception feedback for both sides. Instead of the acknowledgement object the RX can also append any arbitrary response object which may contain already the data requested by TX.

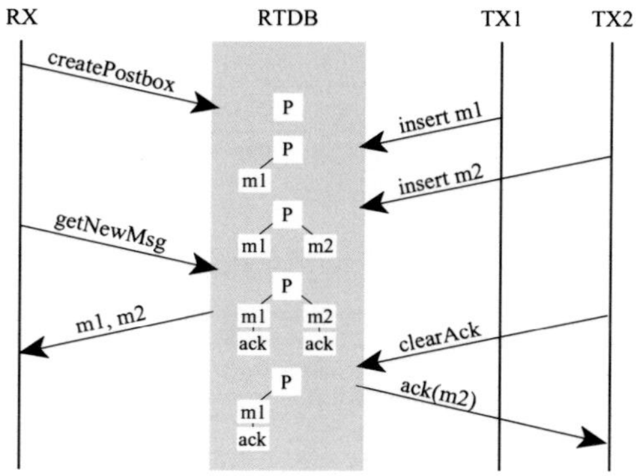

Fig. 2 Scheme of the *ARCADE* postbox system

2.2 Evaluation of the Robot Architecture

In order to further highlight the supported features of *ARCADE* we evaluated it using the conceptual framework for comparing robot architectures introduced in [6]. Other frameworks for architecture comparison were proposed, e.g. in [8]. However, [6] was chosen because of the detailed and specific description of the evaluation criteria and rating method.

Table 1 shows the evaluation criteria and the corresponding level of support in *ARCADE*. A detailed description of all the criteria is given in [6]. As opposed to the original framework we do not consider criteria regarding algorithm implementations such as localization or route planning in our evaluation.

Architectural primitives (F1.1) refer to predefined functional and/or knowledge primitives. *ARCADE* provides fundamental components to control the robotic hardware and to perform a set of basic actions which can be scheduled via a priority-based action controller. However, a kind of generic behavior framework is not provided resulting in a *somewhat supported* mark. With respect to the software engineering criteria (F1.2) *ARCADE* provides coding guidelines and a set of basic classes enabling code reusability and simplification. Nevertheless an explicit theoretical foundation is missing. Architecture neutrality (F1.3) is not provided since the presented framework belongs to the class of blackboard architectures.

Since the RTDB only supports Linux the same applies also to *ARCADE* (F2.1). Additionally besides our own robotic hardware only few further devices – such as the SICK LSM200, S300 or the JR3 force/torque sensor – are currently supported

Table 1 Qualitative evaluation of the *ARCADE* robotic architecture: ⊕ = *well supported*, ⊙ = *somewhat supported*, ⊖ = *not supported*. The criteria codes, names and specifications were taken from [6].

Category	Criteria	*ARCADE*
Specification F1	F1.1 Architectural Primitives	⊙
	F1.2 Software Engineering	⊙
	F1.3 Architecture Neutrality	no
Platform F2	F2.1 Operating System	Linux
	F2.2 Hardware Support	⊖
	F2.3 Simulator	⊙
	F2.4 Configuration Method	⊕
Infrastructure F3	F3.1 Low-level Communication	ICE,RTDB
	F3.2 Logging Facilities	⊕
	F3.3 Debugging Facilities	⊕
	F3.4 Distribution Mechanisms	⊕
	F3.5 Scalability	⊙
	F3.6 Component Mobility	⊙
	F3.7 Monitoring/Management	⊕
	F3.8 Security	⊙
	F3.9 Fault-tolerance	⊙
Implementation F4	F4.1 Programming Language	C++, Simulink
	F4.2 High-level Language	no
	F4.3 Documentation	⊙
	F4.4 Real-time Operation	yes
	F4.5 Graphical Interface	⊙
	F4.6 Software Integration	⊖

(F2.2). Regarding the simulator (F2.3), *ARCADE* makes use of several different simulation/visualization tools, such as the *ARCADE* Visualizer, see Fig. 3, which provide the means for dynamic or multi-robot simulations but are currently separate modules and not yet fully integrated into a single simulator. From the configuration methods point of view (F2.4), XML-files are supported. Furthermore graphical interfaces are provided to set and modify parameters online and to send commands to the respective modules.

The low-level communication (F3.1) through RTDB and ICE was described in section 2.1. The RTDB provides a record and replay functionality with real-time support, i.e. data can be captured of part of or all the database objects and later be replayed keeping the original timing (F3.2). Additionally, console output of any module can be remotely captured and stored in files. Due to the distributed processing, any module – except the central RTDB – can be suspended, modified and restarted during runtime which facilitates the debugging (F3.3). Nevertheless no automatic mechanism keeping care of continuing system operation is provided, which would be required for well supported component mobility (F3.5). As already mentioned, the distribution mechanisms (F3.4) are well supported through the use of ICE and IceStorm. The scalability (F3.5) as defined in [6] is dependent on the values of the other criteria and results in our case in a *somewhat supported* mark.

The *ARCADE Inspector* (Fig. 3) is a sophisticated graphical interface to start/stop modules, view the current state of any database object or send commands providing a convenient tool for monitoring and management (F3.7). Security (F3.8) can only be indirectly supported through usage of the SSL protocol in ICE, and the RTDB support for the single producer concept which allows data modification of an object only by its source process. The distributed processing nature of *ARCADE* makes it fairly tolerant to failures (F3.9), but its notable weaknesses are the dependency of all subscribed modules to their local RTDB and lack of active failure recovery.

ARCADE provides support for C++ and Simulink (F4.1). Even though an interface for high level commands is provided, which could be incorporated into a behavior framework, no explicit high level language is integrated (F4.2). Even though code documentation is available, no API or user guidelines are provided at the moment (F4.3). While none of the compared architectures in [6] provides real-time support (F4.4), it is the main strength and also the crucial motivation for the development of *ARCADE*. While the *ARCADE Inspector* visualizes each module operation, a graphical tool for the design of control code (F4.5) is not provided (except Simulink). A standard API for software integration (F4.6) is *not supported* at all.

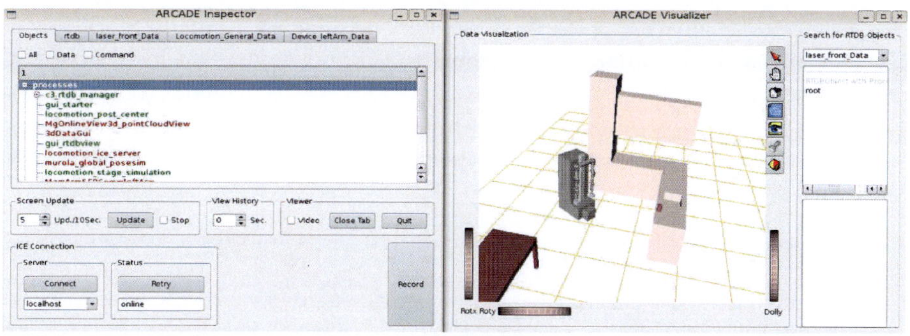

Fig. 3 The *ARCADE* Inspector lists the tree of all database objects, provides access to the object data and a record functionality. The *ARCADE* Visualizer provides a 3D visualization of the current system state.

In order to obtain also a quantitative measure, we calculated the feature score of the reduced criteria list according to [6] using equal weights. For binary-valued criteria a value of 0 or 2, for ternary-valued criteria a value of 0, 1 or 2 is assigned. Accumulating the values in Table 1 leads to a total score of 55%. In a comparison with the architectures in [6] *ARCADE* ranks sixth out of ten. The main weakness of *ARCADE* is the specification category (F1) where it scores 33%, which is below all other architectures. This follows from the initial unavailability of components (F1.1 and F1.2), such as standard hardware drivers, integrated algorithms and clearly structured interfaces which in general are provided by open-source frameworks. The Linux-only support (F2.1) may be regarded a further disadvantage, especially when a large number of researchers are involved, and the need for multi-platform support arises.

ARCADE is very strong in the infrastructure category (F3) achieving a score of 75%, where only *ADE* (100%) ranks better [6]. Consequently, while further work is required in order to simplify its portability, *ARCADE* provides very good tools and mechanisms to develop and operate a running system.

Yet, for highly distributed systems considered in this paper real-time support (F4.4), sophisticated management (F3.7) and distribution mechanisms (F.3.4) are mandatory. As opposed to *ARCADE*, other existing architectures do not adequately meet these aspects.

3 Application Example

In order to illustrate the integration of modules and the usage of the *ARCADE* framework, an example of a fast reactive collision avoidance application for robotic arms based on virtual forces is described. The example application comprises several interconnected modules which are distributed on two computers as shown in Fig. 4, each running an instance of the RTDB. A potential reason for using multiple computers is the distribution of computational load.

The world-builder module (WB) running on computer 1 processes the raw data stream of the sensor module and generates a world model that includes a list of objects and their respective collision shape primitives. This model is mirrored from RTDB 1 to RTDB 2 using the inter-RTDB communication mechanisms.

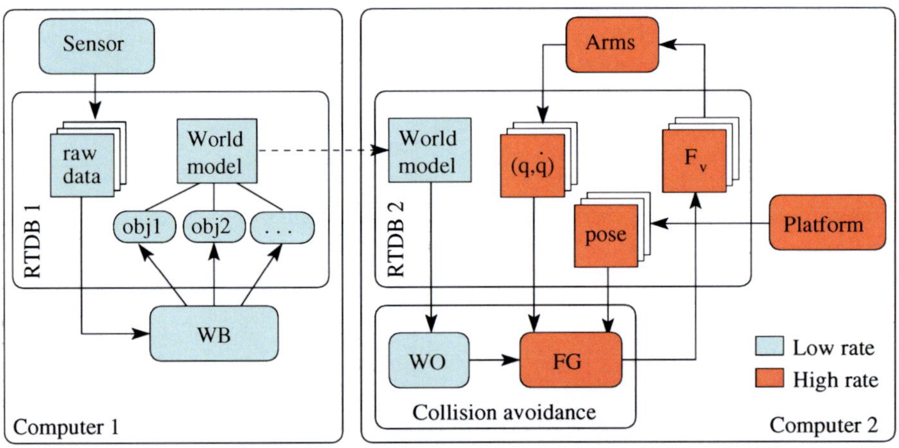

Fig. 4 Overview of the collision avoidance application: The *ARCADE* framework manages distributed computing and different update rates of the modules

The world observer (WO) running on computer 2 monitors changes of the world model and passes a list of relevant world collision shapes to the force generator module (FG). The FG calculates virtual forces F_v acting on the robotic arms based on the current state $(q,\dot{q})$ of the arms and the current pose of the platform.

The update frequency of the world model depends on the kind of sensors and processing algorithms used, but typically is less than 50 Hz. However, the FG is running at the same frequency as the actual control loop, which is typically about 1 kHz.

4 Conclusion

This paper presents *ARCADE*, a data-driven robot architecture combining KogMo-RTDB and the ICE communication middleware. The RTDB provides a central element of *ARCADE* and an easy way for exchange of data between the modules, whereas ICE and IceStorm provide the connection to distributed modules and other RTDBs. The architecture is able to provide both real-time guarantees for low-level robot control and a simple and effective information exchange between distributed modules in multi-robot systems. The presented evaluation showed that *ARCADE* gives a very good solution for distributed multi-robot systems, but still has several weaknesses, mostly due to unavailability of components. The ease of use and versatility of the architecture was illustrated in an application example.

Acknowledgements. We would like to thank Matthias Goebl from the Institute for Real-Time Computer Systems of TU München for allowing us access to the KogMo-RTDB, its documentation and accompanying programs which served as a starting point for our architecture.

This work is supported in part within the DFG excellence initiative research cluster Cognition for Technical Systems - CoTeSys (`www.cotesys.org`).

References

1. Dozio, L., Mantegazza, P.: Real time distributed control systems using RTAI. In: Proceedings of the Sixth IEEE International Symposium on Object-Oriented Real-Time Distributed Computing, Hakodate, Japan (2003)
2. Gerkey, B., Vaughan, R.T., Howard, A.: The player/stage project: Tools for multi-robot and distributed sensor systems. In: Proceedings of the 11th International Conference on Advanced Robotics (ICAR 2003), Coimbra, Portugal, pp. 317–323 (2003)
3. Goebl, M.: Eine realzeitfähige architektur zur integration kognitiver funktionen. Ph.D. thesis, Technische Universität München, Institute for Real-Time Computer Systems (2009)
4. Goebl, M., Färber, G.: A real-time-capable hard- and software architecture for joint image and knowledge processing in cognitive automobiles. In: Proceedings of the 2007 IEEE Intelligent Vehicles Symposium, pp. 734–740 (2007)

5. Henning, M.: A new approach to object-oriented middleware. IEEE Internet Computing 8(1), 66–75 (2004)
6. Kramer, J.F., Scheutz, M.: Development environments for autonomous mobile robots: A survey. Autonomous Robots 22(2), 101–132 (2007)
7. Quigley, M., Conley, K., Gerkey, B., Faust, J., Foote, T.B., Leibs, J., Wheeler, R., Ng, A.Y.: ROS: an open-source Robot Operating System. In: International Conference on Robotics and Automation, Open-Source Software workshop (2009), `http://www.willowgarage.com/`
8. Shakhimardanov, A., Prassler, E.: Comparative evaluation of robotic software integration systems: A case study. In: Proceedings of the 2007 IEEE/RSJ International Conference on Intelligent Robots and Systems, San Diego, CA, USA, pp. 3031–3037 (2007)
9. The Mathworks: Real-Time Workshop 7 - Generate C code from Simulink models and MATLAB code (2007)

Shared-Control Paradigms in Multi-Operator-Single-Robot Teleoperation

Daniela Feth, Binh An Tran, Raphaela Groten, Angelika Peer, and Martin Buss

Abstract. Extending classical bilateral teleoperation systems to multi-user scenarios allows to broaden their capabilities and extend their applicability to more complex manipulation tasks. In this paper a classical Single-Operator-Single-Robot (SOSR) system is extended to a Multi-Operator-Single-Robot (MOSR) architecture. Two shared-control paradigms which enable visual only or visual and haptic coupling of the two human operators are introduced. A pointing task experiment was conducted to evaluate the two control paradigms and to compare them to a classical SOSR system. Results reveal that operators benefit from the collaborative task execution only if haptic interaction between them is enabled.

1 Introduction

This paper extends classical bilateral teleoperation architectures to multi-user scenarios with the aim of broadening their capabilities and extending their applicability to more complex manipulation tasks. The focus of this work is on a Multi-Operator-Single-Robot (MOSR) architecture which enables two humans (H), both operating a human-system interface (HSI) to control collaboratively one teleoperator (TOP) in a shared-control mode.

In literature, MOSR-like shared-control architectures are known from student-teacher scenarios, whereby a trainee is supported by a trainer in performing manipulation tasks. Chebbi et al. [3] introduced three different types of haptic interaction modes suitable for such an scenario: no, unilateral, or bilateral haptic signal exchange. An example for an unilateral information exchange is implemented in [10], whereby a trainee receives position and force feedback about the trainer's actions.

Daniela Feth · Binh An Tran · Raphaela Groten · Angelika Peer · Martin Buss
Institute of Automatic Control Engineering, Technische Universität München, Munich, Germany
e-mail: {daniela.feth,r.groten,angelika.peer,mb}@tum.de

H. Ritter et al. (Eds.): Human Centered Robot Systems, COSMOS 6, pp. 53–62.
springerlink.com

In [4] control architectures that allow a bilateral haptic signal exchange in a virtual bone drilling scenario are presented. Finally, in [8, 11] a dominance factor α is introduced to assign clear roles of dominance to the interacting partners according to their skill level.

Further applications for MOSR systems can be found in the field of semi-autonomous teleoperation. The high physical and cognitive requirements when operating a classical bilateral teleoperation system desire often for a technical assistant which helps the human operator in executing a task [14]. This results in a semi-autonomous teleoperation system that enables the interaction between a human operator and an assistance function and, thus, forms a MOSR system.

Finally, a quite different motivation for MOSR architectures is given by studies of haptic human-human interaction in joint object manipulation tasks. It is shown that task performance of two humans solving a haptic task collaboratively [12] is higher than of a single operator performing the same task. Hence, we expect that adding an additional human operator to a classical bilateral teleoperation scheme has a positive effect on task performance.

Not only single and partner conditions are compared but also the influence of the presence of haptic feedback in technically mediated systems is analyzed. In [2] the performance of a vision-only and vision-haptic condition in a ring-on-wire task is contrasted. They conclude that the interacting partners benefit from the additional haptic feedback. Sallnäss [13] reports the same positive effect of haptic feedback on task performance in a collaborative cube lifting task.

In this paper, we propose a general framework for a MOSR teleoperation system that enables haptic interaction with the remote environment and with the interacting partner. Based on this, we introduce two shared-control paradigms which realize different types of interaction. Unlike teacher-student scenarios, we assume equally skilled human operators and focus on the interaction between them. Furthermore, we consider only tasks which can be performed successfully by a single operator, too. The two paradigms are implemented on a 6 DOF teleoperation system and compared to each other with respect to task performance. Additionally, they are compared to a Single-Operator-Single-Robot teleoperation condition which allows to make statements on the differences between multi-user and single-user teleoperation.

2 Overall MOSR Control Architecture

A classical force-position (F-P) architecture [7] has been extended to form a MOSR system.

The realization of a haptic MOSR system requires to face three main challenges: i) realizing simultaneous control of the remotely located teleoperator, ii) providing feedback from the remote environment, and iii) from the interacting partner.

To approach these challenges, we define three different components: *signal fusion*, *feedback distribution*, and *partner interaction*. Their location within the global

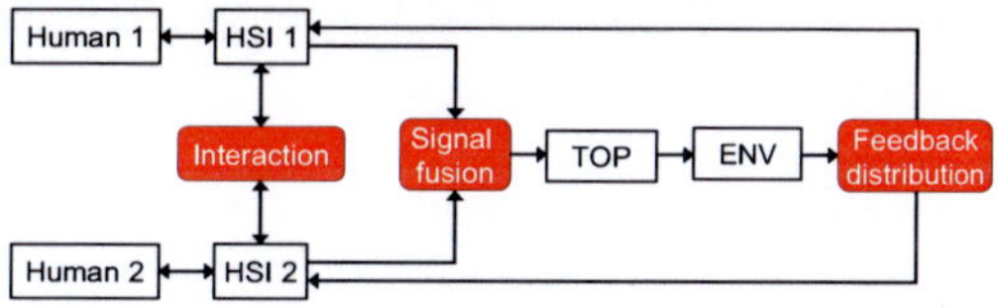

Fig. 1 Multi-Operator-Single-Robot architecture

teleoperation control scheme is illustrated in Fig. 1. Depending on the selected shared-control paradigm different implementations of these components have to be realized.

3 MOSR Shared-Control Paradigms

Two shared-control paradigms enabling different couplings between the human operators and the environment are introduced in this section. They lead to different haptic interaction forms, manifested in different implementations of the above mentioned components *signal fusion*, *partner interaction*, and *feedback distribution*.

3.1 Visual Coupling of Operators

The first shared-control paradigm enables operators to move their haptic interfaces independently. This is achieved by providing only visual, but no haptic feedback about the partner's action and distributing the feedback from the remote environment such that the operator receives only forces caused by her/his own action. In terms of assistance functions, this would correspond to a function that supports the task execution on the remote side and the human operator receives only visual, but no haptic feedback about this support. In order to realize this paradigm the components *signal fusion, partner interaction* and *feedback distribution* are as follows:

Signal fusion: Shared control of the common teleoperator is realized by a weighted sum of the positions of the single haptic interfaces

$$x_m = \alpha x_{m1} + (1-\alpha)x_{m2}. \tag{1}$$

The dominance factor α has been first introduced in [8, 11] to account for different dominance distributions in teacher-student scenarios. We assume both interaction partners to be equally skilled as dominance distributions are not the focus of this work, hence, $\alpha = 0.5$ is chosen.

Partner Interaction: There are no haptic signals exchanged between the haptic interfaces of the interacting operators.

Feedback Distribution: Feedback forces from the remote environment have to be mapped to both of the operators' haptic interfaces. Inspired by [5] we decided for a paradigm where each human operator receives only haptic feedback about her/his own action. Thus, the operator receives force feedback only if she/he really contributes to the interaction force f_{env} with the environment, which means that f_{env} points to the opposite direction of f_h, the force applied by the respective operator. Force feedback is divided as follows: If the partner does not apply a force against f_{env}, the operator receives full force feedback from the environment $f_{env1} = f_{env}$. However, if both partners apply a force against f_{env}, force feedback is distributed as follows:

$$f_{env1} = \beta f_{env}, \qquad f_{env2} = (1-\beta) f_{env}, \qquad \text{with} \quad \beta = \frac{f_{h1}}{f_{h1} + f_{h2}}. \tag{2}$$

This kind of shared-control paradigm can not be realized in real physical scenarios, but in teleoperation or in virtual reality. It enables free-space motion of the operators without any interference by the partner as they act independently. But, the human operators can infer on their partner's action only by the visual motion of the teleoperator. Haptic and visual feedback are inconsistent.

3.2 Visual and Haptic Coupling of Operators

The second shared-control paradigm realizes a visual and haptic coupling of the operators, see Fig. 2. Imagine two human operators holding the ends of a rigid object,

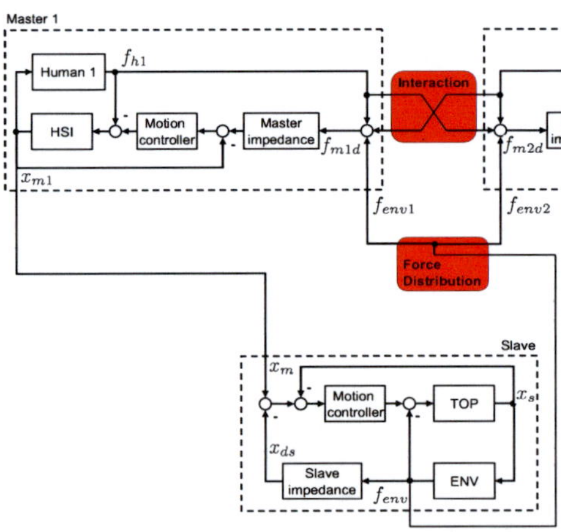

Fig. 2 Teleoperation architecture with local position-based admittance controllers applied for visual and haptic coupling of operators

e.g. a rod, and interacting with the environment via this rod. In our teleoperation scenario, the rod is represented by the teleoperator and the required coupling of the operators is realized by a virtual rigid connection between the two haptic interfaces. Again, the components to achieve such a coupling are discussed in the following paragraphs.

Signal Fusion: The stiff coupling of the operators causes $x_{m1} = x_{m2} = x_m$. Hence, only the position of one haptic interface has to be sent to the remotely located teleoperator.

Partner Interaction & Feedback Distribution: Due to the rigid connection of the operators they receive full haptic feedback from their partner as well as from the remote environment. Hence, the desired forces f_{m1d} and f_{m2d} to be displayed by the haptic interfaces are calculated by

$$f_{m1d} = f_{m2d} = f_{h1} + f_{env} + f_{h2} \qquad \text{where} \qquad f_{env} = f_{env1} = f_{env2}. \tag{3}$$

In this shared-control paradigm visual and haptic feedback are congruent without any inconsistencies. Operators receive full information about the task execution and the partner's behavior. However, the operators' actions are not independent as they strongly depend on the partner's behavior what results in a restricted operating range.

4 Teleoperation System

Both shared-control algorithms are implemented on a real teleoperation system with visual and haptic feedback devices. As shown in Fig. 3(a) the operator side consists of two admittance-type haptic input devices with 6 DOF. The kinematics and technical specifications of the teleoperator (Fig. 3(b)) are the same as the ones of the haptic input devices. A 6 DOF force/torque-sensors (JR3) is mounted at the tip of the manipulators to measure interaction forces with the human operator and the environment. For details on the teleoperation system please refer to [1, 9].

The control of the telemanipulation system, see Fig. 2, was implemented in Matlab/Simulink. Real-time capable code is generated by using the Matlab Real-Time Workshop and executed on two computers with the Linux Real-Time Application Interface (RTAI). Communication between the two computers is realized by an UDP connection in a local area network, thus time delay can be neglected.

Visual feedback was provided by two computer monitors displaying the image of a CCD firewire camera that captures the remote environment and teleoperator. Thus, both operators had the same fixed view point and, hence, the same information about the remote environment.

5 Experimental Evaluation

An experiment was conducted to evaluate the implemented MOSR shared-control paradigms with respect to task performance. We compared the two implemented control strategies of operator coupling to a classical bilateral teleoperation system where one human operates one teleoperator. As an exemplary task a pointing task was chosen, requiring fast as well as accurate behavior of the operators.

Task: Participants carried out a pointing task by controlling their haptic input devices such that the teleoperator's end-effector moved from one desired position to the next. These target positions were visualized as 4 circles with a radius of 2 cm on a sheet of paper which was placed underneath the end-effector of the teleoperator, see Fig. 3(b) and Fig. 4(a).

Participants were instructed to move the tip of a pen mounted on the teleoperator's end-effector as fast and as accurate as possible from a starting position to the center of one of the circles. By touching the surface of the paper with the pen a dot was drawn on the paper to provide visual feedback to the operators. A target

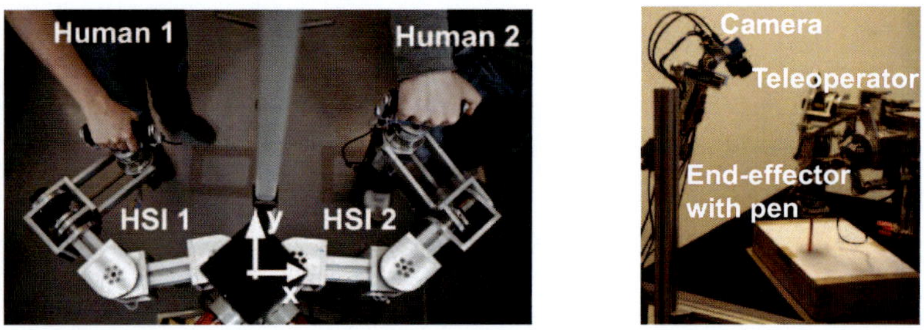

(a) Haptic input devices on operator side (b) Teleoperator side

Fig. 3 6 DOF teleoperation system

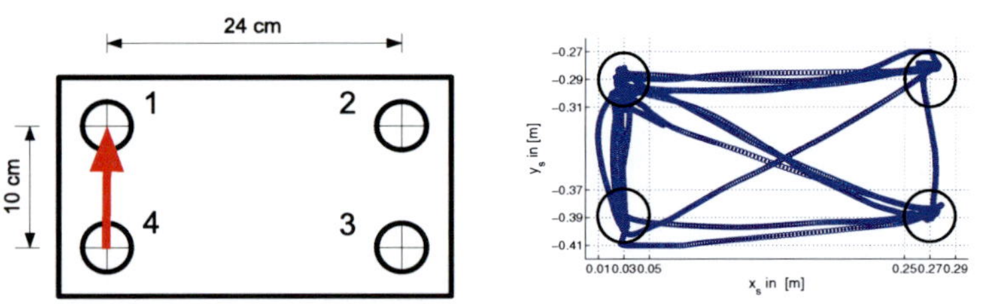

(a) Target positions indicated by circles; the arrow indicates the next target

(b) x-y position of one trial (VH conditon)

Fig. 4 Pointing task

was achieved successfully as soon as the surface was contacted by the pen within the desired circle (if $F_z > 1$ N). After the first target was achieved a new circle was assigned as next target. Information about the current target circle was provided by a second monitor placed next to the camera image. The current step was always marked with an arrow as indicated in Fig. 4(a).

The order of the targets was a random combination of four pre-defined paths each containing four different circles (e.g. $1 \rightarrow 2 \rightarrow 3 \rightarrow 4$). The length of the resulting overall paths was kept constant.

Experimental Design & Procedure: Three experimental conditions were analyzed. The already introduced shared-control paradigms were contrasted to a single condition, whereby the meaning of each experimental condition can be summarized as follows:

- Single (S): each participant performed the task alone
- Visual coupling (V): operators were only visually coupled
- Visual and haptic coupling (VH): operators were visually and haptically coupled

In the experiment, 26 participants (13 female/ 13 male, age $= 25.04 \pm 2.79$ years) took part. They were assigned pseudorandomly to 13 independent couples of 1 male and 1 female. Each participant performed the task once in each of the three randomized conditions.

Data Analysis: We analyzed task performance by evaluating the task error (accuracy) and task completion time (speed).

The *task error* ε of each trial is defined by the mean Euclidian distance of the dot's position $(x_{dot}|y_{dot})$ (drawn by the participants) from the center of the respective target circle $(x_c|y_c)$

$$\varepsilon = \frac{1}{M}\sum_{i=1}^{M}\sqrt{(x_c - x_{dot})^2 + (y_c - y_{dot})^2} \tag{4}$$

with $M = 16$ the total number of target circles in each trial. If multiple points were drawn, the first point inside the circle was taken.

The *task completion time* tct is defined as the time required to perform the task successfully.

Results: We analyzed the experimental data with respect to the above introduced performance measures for each of the three experimental conditions. The x-y teleoperator position for an exemplary VH trial is shown in Fig. 4(b). Table 1 as well as Fig. 5(a) and 5(b) show the means and standard deviations.

As can be observed there is no difference in task performance with respect to the *mean task error* for each of the three conditions (one-factorial repeated measurement ANOVA, Greenhouse-Geisser corrected, $F(1.3, 15.60) = 0.994$, $p = 0.357$).

Task completion time is smallest in the VH condition. A one-factorial repeated measurement ANOVA (Greenhouse-Geisser corrected, $F(1.16, 13.93) = 14.86$,

Table 1 Experimental results: Means and std. deviations of performance measures

	$\overline{\varepsilon}$ in [m]	σ_ε	$\overline{tct}$ in [s]	σ_{tct}
S	0.0069	0.0017	72.64	24.99
V	0.0076	0.0014	54.11	16.43
VH	0.0078	0.0023	41.16	9.51

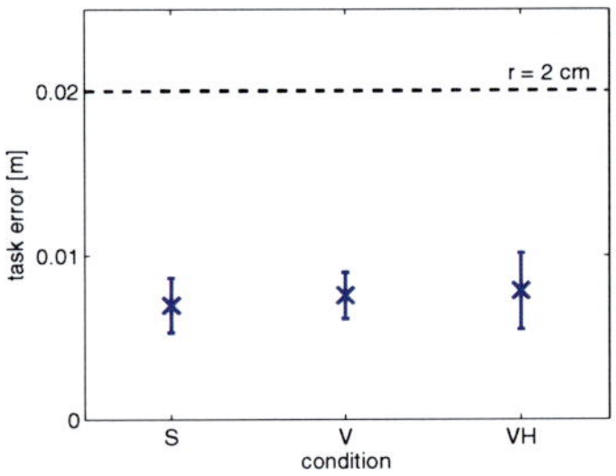

(a) Task error (mean and standard deviation); $r = 2$ cm is the radius of the target circles

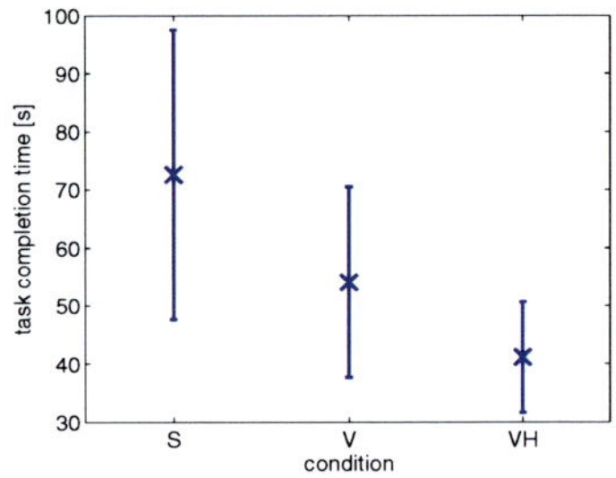

(b) Task completion time (mean and standard deviation)

Fig. 5 Task performance in the three experimental conditions

$p = 0.001$, partial $\eta^2 = 0.553$) reveals a significant effect of the shared-control conditions on task completion time. Post-hoc Bonferroni adjusted pairwise comparisons show a significantly better task completion time in the VH condition compared to the S as well as the V condition. There is no significant difference between S and V.

Discussion: For interpretation of the achieved results our special task design has to be considered: Participants could move to the next target only if they placed the pen inside the current target circle which automatically causes the task error to be upper bounded by 2 cm. This explains why the task completion time varies between the conditions, but the task error remains approximately constant.

In the here presented teleoperation scenario task performance is better if haptic feedback is provided compared to the V and S condition. This is consistent with results reported by related work out of the field of haptic human-human interaction [2, 13, 12].

However, our results reveal no significant difference of task performance in S and V condition. This is contradictory to the results we presented on a 1 DOF pursuit tracking task in [6]. We assume this is due to the fact that the multi-DOF pointing task is more complex and requires higher motion coordination of the operators than the 1 DOF tracking task. In fact, during the experiment we observed that participants had difficulties to synchronize their actions in the V condition.

6 Conclusion

In a Multi-Operator-Single-Teleoperator scenario appropriate shared-control laws have to be defined to enable i) *fusion of the signals sent to the teleoperator*, ii) *distribution of the feedback received from the remote environment*, and iii) *haptic interaction between the human operators*. We introduced two different Multi-Operator-Single-Teleoperator (MOSR) shared-control paradigms that are characterized by a visual coupling only and visual and haptic coupling of the operators. According to the desired behavior of the MOSR teleoperation system the three parts i)-iii) were defined and have been realized on a 6 DOF teleoperation system. The achievable task performance using these couplings was evaluated by an experiment comparing them to a classical SOSR teleoperation system. Results showed that adding a second human operator had only a positive effect on task performance if haptic feedback of the partner was provided. The results cannot be compared to findings obtained in student-teacher scenarios [8, 11] due to their different dominance distributions. Furthermore, the observed benefit might be task-dependent. Hence, further studies are needed for generalization.

Future research will focus on defining further MOSR shared-control architectures to improve task performance in haptic interaction tasks as well as the stability analysis of the proposed architectures.

Acknowledgements. This work is supported in part by the German Research Foundation (DFG) within the collaborative research center SFB453 'High- Fidelity Telepresence and Teleaction'.

References

1. Baier, H., Buss, M., Freyberger, F., Hoogen, J., Kammermeier, P., Schmidt, G.: Distributed PC-based haptic, visual and acoustic telepresence system-experiments in virtual and remote environments. In: Proceedings of the IEEE Virtual Reality, pp. 118–125 (1999), doi:10.1109/VR.1999.756942
2. Basdogan, C., Ho, C.H., Srinivasan, M.A., Slater, M.: An experimental study on the role of touch in shared virtual environments. ACM Trans. Comput.-Hum. Interact. 7(4), 443–460 (2000), `http://doi.acm.org/10.1145/365058.365082`
3. Chebbi, B., Lazaroff, D., Bogsany, F., Liu, P., Ni, L., Rossi, M.: Design and implementation of a collaborative virtual haptic surgical training system. In: IEEE International Conference on Mechatronics and Automation, vol. 1, pp. 315–320 (2005)
4. Esen, H.: Training in virtual environments via a hybrid dynamic trainer model. Ph.D. thesis, Technische Universität München (2007)
5. Glynn, S.J., Fekieta, R., Henning, R.A.: Use of force-feedback joysticks to promote teamwork in virtual teleoperation. In: Proceedings of the Human Factors and Ergonomic Society 45th Annual Meeting (2001)
6. Groten, R., Feth, D., Klatzky, R., Peer, A., Buss, M.: Efficiency analysis in a collaborative task with reciprocal haptic feedback. In: The 2009 IEEE/RSJ International Conference on Intelligent Robots and Systems (2009)

7. Hashtrudi-Zaad, K., Salcudean, S.E.: Analysis of control architectures for teleoperation systems with impedance/admittance master and slave manipulators. The International Journal of Robotics Research 20, 419–445 (2001)
8. Khademian, B., Hashtrudi-Zaad, K.: A four-channel multilateral shared control architecture for dual-user teleoperation systems. In: IEEE/RSJ International Conference on Intelligent Robots and Systems, IROS 2007, pp. 2660–2666 (2007), doi:10.1109/IROS.2007.4399225
9. Kron, A., Schmidt, G.: Haptic telepresent control technology applied to disposal of explosive ordnances: Principles and experimental results. In: Proceedings of the IEEE International Symposium on Industrial Electronics, ISIE 2005, pp. 1505–1510 (2005)
10. Morris, D., Sewell, C., Barbagli, F., Salisbury, K., Blevins, N., Girod, S.: Visuohaptic simulation of bone surgery for training and evaluation. IEEE Computer Graphics and Applications 26(6), 48–57 (2006)
11. Nudehi, S., Mukherjee, R., Ghodoussi, M.: A shared-control approach to haptic interface design for minimally invasive telesurgical training. IEEE Transactions on Control Systems Technology 13(4), 588–592 (2005)
12. Reed, K.B., Peshkin, M.A.: Physical collaboration of human-human and human-robot teams. IEEE Transactions on Haptics 1(2), 108–120 (2008)
13. Sallnäs, E.L.: Improved precision in mediated collaborative manipulation of objects by haptic force feedback. In: Brewster, S., Murray-Smith, R. (eds.) Haptic HCI 2000. LNCS, vol. 2058, pp. 69–75. Springer, Heidelberg (2001)
14. Weber, C., Nitsch, V., Unterhinninghofen, U., Färber, B., Buss, M.: Position and force augmentation in a telepresence system and their effects on perceived realism. In: Third Joint Eurohaptics Conference and Symposium on Haptic Interfaces for Virtual Environment and Teleoperator Systems (2009)

Assessment of a Tangible User Interface for an Affordable Humanoid Robot

Jacopo Aleotti and Stefano Caselli

Abstract. The paper reports a systematic evaluation of a tangible user interface (TUI) for programming robot tasks in the context of humanoid robotics. The assessment is aimed at exploring the potential benefits of a natural tangible interface for human-robot interaction (HRI) compared to traditional keypad remote controllers. The proposed user interface exploits the Nintendo Wii® wireless game controller whose command signals have been used to drive a humanoid platform. An affordable robot (the Robosapien V2) has been adopted for the experimental evaluation. Two different tasks have been considered. The first experiment is a walking task with obstacle avoidance including object kicking. The second experiment is a gesture reproduction task, which involves both arms and upper body motion. The gesture based TUI has proven to decrease task completion time.

1 Introduction

Advanced humanoid robots are expected to work as service machines to assist people in daily life activities. However, programming such complex machines is a difficult and demanding challenge. In this paper we focus on the problem of humanoid remote operation. When a user requires his/her personal robot to perform complex or highly specific tasks he/she may require a low level of autonomy. Hence, if a high degree of human intervention is desired the design of an appropriate user interface becomes essential. It is a widely accepted concept that traditional user interfaces, such as mouse and keypads are unsuitable for complex HRI tasks. The reason is that such devices do not offer an adequate physical embodiment, meaning that it is not straightforward to understand the correct mapping between the input devices and the resulting actions. The absence of a physical meaning associated to standard

Jacopo Aleotti · Stefano Caselli
Dipartimento di Ingegneria dell'Informazione, University of Parma, Italy
e-mail: {aleotti,caselli}@ce.unipr.it

H. Ritter et al. (Eds.): Human Centered Robot Systems, COSMOS 6, pp. 63–72.
springerlink.com

control devices makes them unreliable for memorization and potentially more time consuming to learn. Moreover, traditional user interfaces decouple the two phases of action and perception since they require a periodic switch of attention between the device and the observed environment where the robot acts.

Tangible interfaces can overcome the disadvantages of traditional devices by providing more natural ways for operating a robot. However, commercial top-end TUIs often come at a high cost. In our previous work [1, 2] we have investigated the capabilities of the affordable humanoid robot RoboSapien V2 (RSV2). In particular, we have considered both visual-guided walking tasks and gesture mimicking. The reliability of the proposed approach was strongly dependent on the accurate motion tracking devices that were adopted. In this work our goal is to evaluate human-robot interaction tasks relying on a fully affordable humanoid platform. To this purpose we have chosen the Nintendo Wiimote controller, a low-cost gestural TUI suitable for motion tracking. We have carried out two rather complex HRI experiments and we provide a quantitative evaluation of the two tasks. Two interfaces have been considered for comparison, namely, the gestural Wiimote interface and a non-gestural TUI based on a joypad. Few authors have considered the use of the Robosapien in past works. Behnke et al. [3] have investigated the capabilities of the Robosapien V1 robot for playing soccer. The Nintendo Wiimote device has been investigated in different works in the context of human-robot interaction. Gams and Mudry [4] have successfully applied the Wiimote for controlling a HOAP-2 humanoid robot in a drumming task. Guo and Sharlin [5] have used the Wiimote as a tangible interface for controlling the AIBO robot dog. A user study has been performed to compare the Wiimote interface with a traditional keypad device. Results provided evidence that a gesture interface can outperform traditional input modalities. In [7] further demonstrations of the Wii controller are presented in HRI tasks for both the Sony AIBO and the iRobot Roomba. In [13] an advanced interface device for HRI has been developed. The interface includes the Nintendo Wiimote and improves its accuracy by compensating errors in acceleration measures with an additional force sensor. Varcholik et al. [15] presented a qualitative evaluation of the Wiimote device as a tool for controlling unmanned robot systems to improve usability and to reduce training costs. Shiratori and Hodgins [11] proposed a user interface for the control of a physically simulated character where two Wiimotes are used to detect frequency and phase of lower legs motions.

Hereafter we briefly review further contributions that have addressed HRI tasks by exploiting natural gesture and tangible interfaces. Hwang et al. [6] proposed a fingertip-based method for the generation of mobile robot trajectories. In [8] a haptic system has been developed for mobile robot teleoperation using force feedback. The haptic feedback improved both navigational safety and performance. Moon et al. [9] proposed an intelligent robotic wheelchair with a user-friendly interface including electromyogram signals, face directional gestures, and voice. Sato et al. [10] focused on a pointing interface for HRI tasks. The system interacts with the user and behaves appropriately in response to user's pointing actions. Singh et al. [12] implemented a system where the AIBO robot performs behaviors in response to user's gestures that are acquired and recognized using real-time monocular vision. In [14] a humanoid platform is presented which enables natural human interaction with several components such as speech recognition, people detection and gesture recognition.

2 Affordable Humanoid Platform

The humanoid robot that has been adopted in this work is a Robosapien V2 (RSV2) developed by WowWee and shown in figure 1. RSV2 is an affordable robot expressly designed for the consumer market. It has 60cm height, 7Kg weight, 12 degrees of the freedom and it is driven by 12 DC motors powered by 10 batteries. The robot has realistic joint movements and its functionalities include true bi-pedal walking with multiple gaits, turning, bending, kicking, sitting and getting up. RSV2 can also pick up, drop and throw small objects with articulated fingers. Walking is achieved by alternatively tilting the upper body of the robot and activating the leg motors in opposite directions. The lateral swinging motion of the upper body generates a periodic displacement of the center of mass between the two feet. There are built-in sensors such as an infrared vision system, a color camera, stereo sonic sensors and touch-sensitive sensors in his hands and feet. Signals collected by the sensors can be used to trigger simple reactive behaviors to environmental stimuli through preprogrammed motion sequences, which can be stored in the onboard memory. Sensor information has not been considered in this work.

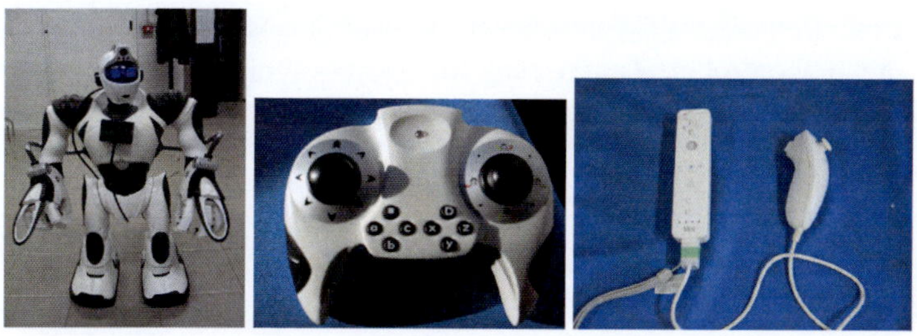

Fig. 1 From left to right: Robosapien V2 (RSV2), RSV2 remote controller, Wiimote and Nunchuk

RSV2 is fully controllable by a remote infrared controller which has been modeled after a video game joypad. The remote controller is shown in figure 1. This device is a non-gestural TUI that has been adopted in the experimental evaluation for comparison with the gestural TUI described below. The controller comprises eight buttons on the top and three shift buttons on the front side. There are also two sticks with eight positions each that are used as follows. The left stick controls the robot's forward, backwards, turn left and turn right bipedal walking. The right stick controls the hands, arms, waist, torso and head movements when combined with the shift keys. Other commands such as kicking or upper body motions require the

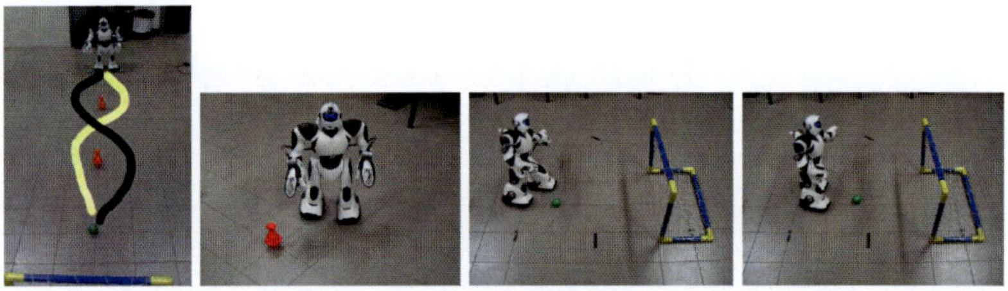

Fig. 2 Obstacle course and images of a navigation task

user to press two buttons at the same time as well as one of the sticks. The remote controller also has a tri-color LED that confirms which shift keys are being pressed. It is a common opinion that learning the signals of the Robosapien V2 is a time consuming job due to the large repertoire of movements and the unnatural control method. Complete mastery of the suite of movements via the remote control may be a daunting task. A common frustrating occurrence is that while driving the robot the user has to switch to the manual to read how to perform a command.

An alternative method to control the RSV2 is to bypass the remote controller with an infrared transmitter connected to a host PC. IR codes of the RSV2 remote controller have been decoded and stored in a database. The infrared transmitter has been mounted on the torso of the robot (Figure 1) close to the RSV2 native infrared receiver, thus helping transmission of command signals. This approach enabled the design of the gestural tangible user interface (TUI) proposed in this paper. The TUI exploits the Nintendo Wii controller. Processing of user's input works as follows. The Nintendo controller is connected via bluetooth to the host PC (compliant with the HID standard). Command signals generated by moving the Wiimote are interpreted and converted into the corresponding infrared codes. Finally, infrared codes are transmitted to the robot. The Wiimote is an inertial control interface that is able to provide two rotational degrees of freedom. Measure of rotational displacement occurs with the help of three accelerometers comprised in a ADXL330 chip from Analog Devices. The Wiimote provides a resolution of $0.23m/s^2$ on a $\pm 3g$ range on each axis (8 bits). The sampling rate is about $100Hz$. Ten buttons of the Wiimote have not been used in the tests. The presence of unused buttons on the Wiimote balances all the unused buttons on the gamepad. The Wiimote is enhanced with an extension for two-handed user input called Nunchuk (also shown in Figure 1). The Nunchuk adds a second set of three accelerometers along with an analog joystick and two buttons. A detailed description of the mapping of both interfaces for the two tasks is provided in section 4 and 5.

3 Experimental Method and Participants

In this section we report the methodology that has been adopted for the empirical investigation and the composition of the subjects that participated in the two experiments. For both tasks 20 individuals were recruited among students at the University of Parma. The age of the subjects varied between 22 and 30 years (mean age was 26 years). Only 4 participants had used the Nintendo Wiimote before. About 50% of the participants declared themselves acquainted with joypads in computer games. 90% of the subjects were right-handed. For each task the users performed a practice session before the actual test with both interfaces. Half of the subjects performed their test (gesture reproduction or navigation) with the Wiimote interface (gestural TUI) first and then a second run with the joypad (non-gestural TUI). The second half performed their test in reverse order. All the participants signed a consent form and answered a questionnaire with the purpose of collecting a qualitative evaluation of the proposed HRI interfaces (lowest score is 1, highest score is 5).

4 Navigation Task

The first experiment that is reported in this paper is a navigation task. Users were asked to drive RSV2 through an obstacle course. The obstacle course (about $250cm \times 150cm$) is shown in Figure 2. It comprises two obstacles and a small soccer net. The robot has to walk around the obstacles by following one of the two alternative paths highlighted in Figure 2. Then RSV2 has to kick a green ball towards the soccer net. Timer is stopped once the ball enters the soccer net. In order to perform the kick the robot has to be placed in a correct position and orientation with respect to the ball. If the robot does not score a goal then the ball is replaced at the initial position and the user has to retry the kicking action. Figure 2 also shows images taken from one of the experiments. At the end of the experiment each user scored the usability of the two interfaces.

Table 1 Navigation task: input commands

Command	Wiimote gestural TUI	Joypad non-gestural TUI
Turn right	Wiimote Roll RIGHT	left stick RIGHT
Turn left	Wiimote Roll LEFT	left stick LEFT
Walk forward	Wiimote Pitch UP	left stick UP
Walk backward	Wiimote Pitch DOWN	left stick DOWN
Stop	Wiimote horizontal rest	stop button
Right kick	Nunchuk Roll RIGHT	SH3+A button
Left kick	Nunchuk Roll LEFT	SH3+Z button

Table 1 illustrates the input commands required to drive the robot. In this experiment the left stick of the joypad (non-gestural TUI) is used to move the robot and a combination of two keys is used for kicking. The Wii-based interface (gestural TUI) exploits both the Wiimote and Nunchuk. In order to drive the robot forward and backward the user has to pitch the Wiimote. Rolling the Wiimote left and right

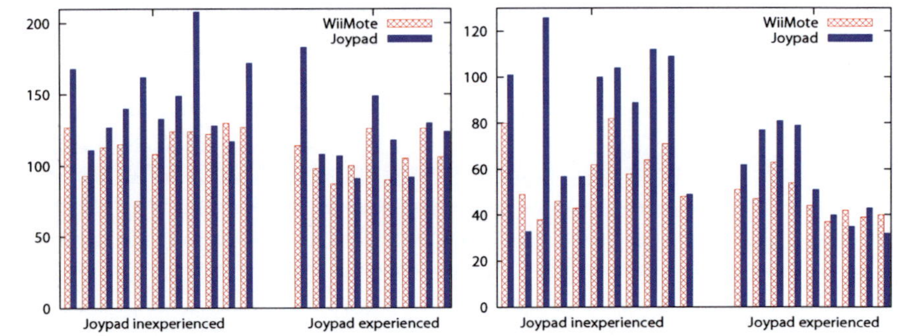

Fig. 3 Completion time for navigation task (left image) and completion time for gesture reproduction task (right image)

Table 2 Navigation task: overall results

	Wiimote gestural TUI		Joypad non-gestural TUI	
	Mean(sec)	Std. dev.	Mean(sec)	Std. dev.
All users	110.5	15.9	135.85	30.76
Experienced users	105.77	14.06	122.44	29.18
Inexperienced users	114.36	16.90	146.81	28.66
Kicking errors	6		11	
User score (1-5)	3.95	0.89	2.60	1.19

make the robot turn. The Nunchuk is used to perform left or right kicks. Figure 3 (left image) shows a bar chart of the completion time for the task, while Table 2 reports the overall results. The mean completion time for the Wiimote gestural interface (110.5*s*) was 18.6% lower than the joypad (135.85*s*). The standard deviation of the gestural TUI (15.9*s*) is half the value of the non-gestural TUI (30.76*s*), suggesting that the Wiimote interface is easier to learn. An ANOVA analysis of variance showed that the difference is statistically significant with $p < 0.02$. Moreover, the total number of kicking errors for the Wiimote interface (6) was notably lower than the total number of errors for the joypad interface (11).

For the purpose of post hoc analysis users were divided into two classes according to their stated degree of expertise in joypad usage for computer games (11 inexperienced users and 9 experienced users). The best performance with the Wiimote interface (75*s*) was achieved by an inexperienced user who got one of the worst completion time with the joypad (162*s*), thus confirming the observation that the proposed tangible interface is more accurate and easier to learn. The best performance with the joypad interface was achieved by an experienced user (91*s*) who also got a good performance with the Wiimote (below the mean value). Only three users performed better with the joypad interface than with the Wiimote. All inexperienced users got better results with the Wiimote interface (114.36*s* mean) than with the joypad (146.81*s* mean).

Users having high confidence with joypads had the same behavior. Indeed, they got a mean completion time of $122.44s$ with the joypad and $105.77s$ with the Wiimote. From the ANOVA test the level of expertise was only marginally significant as discriminant factor between the two groups of users ($p < 0.08$) for the joypad interface while it was not significant for the Wiimote interface. Subjective evaluation of the two interfaces confirmed the quantitative evaluation as the Wiimote interface got a score of 3.95 against 2.60 for the joypad interface.

5 Gesture Reproduction Task

A different group of users was involved in a second task which is reported in this section. The task consists in a gesture reproduction experiment. Users were asked to drive the upper body and the arms of the robot to guide RSV2 through a predefined sequence of postures with both interfaces. Figure 4 shows the sequence of seven postures that were shown to the users while performing the task. Figure 4 also shows a user performing the task. Users were free to choose whether to stay in front of the robot or at its side. Images representing next postures to be performed were shown to the user only after the correct reproduction of the previous gestures in the sequence. At the end of the experiment each user scored the usability of the two interfaces. The sequence of actions are the following: 1 initial configuration (standing with both arms down), 2 raise right arm, 3 lower right arm and lower upper body, 4 raise left hand, 5 raise upper body and raise both arms, 6 lower right arm, 7 lower left arm (back to initial configuration).

Table 3 reports the input commands required to drive the robot. In this experiment the joypad interface required even more efforts since many command require the combined pressure of multiple buttons. On the contrary, the Wiimote interface was easier

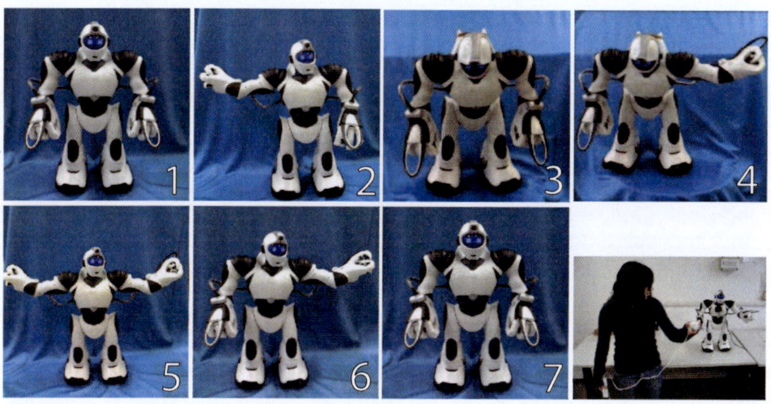

Fig. 4 Sequence of postures for the gesture reproduction task and a picture of the experimental setup

Table 3 Gesture reproduction task: input commands

Gesture	Wiimote gestural TUI	Joypad non-gestural TUI
Right arm UP	Wiimote UP	SH1+right stick UP
Right arm DOWN	Wiimote DOWN	SH1+right stick DOWN
Left arm UP	Nunchuk UP	SH2+right stick UP
Left arm DOWN	Nunchuk DOWN	SH2+right stick DOWN
Upper body UP	Wiimote up arrow	SH1+SH2+SH3+r_stick UP
Upper body DOWN	Wiimote down arrow	SH1+SH2+SH3+r_stick DOWN

Table 4 Gesture reproduction task: overall results

	Wiimote gestural TUI		Joypad non-gestural TUI	
	Mean(sec)	Std. dev.	Mean(sec)	Std. dev.
All users	52.9	13.6	71.9	29.8
Experienced users	46.33	8.39	55.55	19.67
Inexperienced users	58.27	14.93	85.18	30.68
User score (1-5)	4.05	0.69	3.35	0.88

to learn as confirmed by the experimental results. Users had simply to swing the Wiimote (or the Nunchuk) to rotate the arms. The upper body motion has been mapped to the up and down arrows keys of the Wiimote. Figure 3 (right image) shows a bar chart of the completion time for the task, while Table 4 reports the overall results. The mean completion time for the Wiimote interface (52.9s) was 26% lower than the joypad (71.9s). The difference is statistically significant with $p < 0.02$. The standard deviation for the gestural TUI is again considerably lower. The main advantage of the Wiimote interface again stems from the reduced mental overhead required to learn and reproduce the correct command sequence. As in the previous experiment users were divided into two classes according to their level of expertise on joypad usage for computer games. The analysis of the individual results provided more interesting insights. The second best performance with the Wiimote interface was achieved by an unexperienced user. The subject that achieved the best performance with the joypad interface is an experienced user. He declared that the Wiimote interface was usable as well. There were only three users that performed better with the joypad interface than with the Wiimote. It turns out that in this experiment the degree of expertise was a statistically significant factor to discriminate between the two groups of users with both interfaces ($p < 0.05$). All inexperienced users performed better with the Wiimote interface (58.27s mean) than with the joypad (85.18s mean). Users having high confidence with joypads had the same behavior on average, showing a mean completion time of 46.33s with the Wiimote and 55.55s with the joypad. Subjective scores confirmed the quantitative evaluation as the Wiimote interface got a mean of 4.05 points against 3.35 points for the joypad interface.

6 User Learning Rate

A final experiment has been performed to assess the learning rate of the two interfaces. Four users have been asked to repeat the gesture reproduction task for three times. Experiments have been repeated at intervals of two hours. The sequence of postures has been changed between the sessions. Figure 5 shows a graph that summarizes the mean completion time. Results indicate that the mean time decreases as more trials are carried out for both interfaces. In particular, the performance with the Wiimote interface is constantly better. However, the gap between the two interfaces decreases and after three trials users are able to achieve almost the same results with both interfaces. The mean completion time for the joypad interface decreases from 84.25*s* in the first attempt to 34.25*s* in the third, whereas the mean time for the Wiimote interface decreases from 59.75*s* in the first trial to 32*s* in the third. This experiment suggests that training is important to learn the correct use of both interfaces and confirms that the Wiimote interface is less demanding and more effective. When the user has become more skilled, the completion time is dominated by the physical time taken by the humanoid to reproduce the task, and hence the interface plays a lesser role.

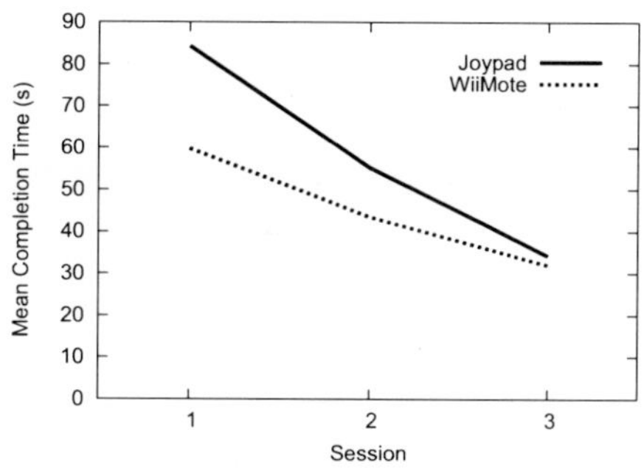

Fig. 5 Learning experiment

7 Conclusions

In this paper we have presented an experimental evaluation of a gestural tangible user interface in programming tasks for a low-cost humanoid robot. The Robosapien V2 has been chosen because it is a promising low-cost robot which provides interesting human-like capabilities in spite of its kinematics limitations. The proposed interface relies on the Nintendo Wii® wireless controller, which has proven easy-to use and more effective than traditional keypad controllers in terms of completion time and success rate. Novice users have found themselves more comfortable by using the TUI, users that have strong experiences in using standard game-pad controllers got better results by exploiting the newly propose interface. The main advantage stems from the reduced overhead required to learn the usage of the novel interface.

References

1. Aleotti, J., Caselli, S.: Imitating Walking Paths with a Low-Cost Humanoid Robot. In: Int'l. Conference on Advanced Robotics, ICAR 2007, Jeju, Korea (2007)
2. Aleotti, J., Caselli, S.: A Low-Cost Humanoid Robot with Human Gestures Imitation Capabilities. In: IEEE Int'l. Symposium on Robot and Human Interactive Communication, RO-MAN 2007, Jeju, Korea (2007)
3. Behnke, S., Müller, J., Schreiber, M.: Playing soccer with robosapien. In: Bredenfeld, A., Jacoff, A., Noda, I., Takahashi, Y. (eds.) RoboCup 2005. LNCS (LNAI), vol. 4020, pp. 36–48. Springer, Heidelberg (2006)
4. Gams, A., Mudry, P.A.: Gaming controllers for research robots: controlling a humanoid robot using a wiimote. In: Int'l. Electrotechnical and Computer Science Conference, ERK 2008 (2008)
5. Guo, C., Sharlin, E.: Exploring the Use of Tangible User Interfaces for Human-Robot Interaction: A Comparative Study. In: SIGCHI conference on Human factors in computing systems (2008)
6. Hwang, J.H., Arkin, R.C., Kwon, D.S.: Mobile robots at your fingertip: Bezier curve on-line trajectory generation for supervisory control. In: IEEE/RSJ Int'l. Conference on Intelligent Robots and Systems (IROS), Las Vegas, Nevada (2003)
7. Lapping-Carr, M., Jenkins, O., Grollman, D., Schwertfeger, J., Hinkle, T.: Wiimote Interfaces for Lifelong Robot Learning. In: AAAI Symposium on Using AI to Motivate Greater Participation in Computer Science, Palo Alto, CA, USA (2008)
8. Lee, S., Sukhatme, G., Jounghyun Kim, G., Park, C.M.: Haptic Teleoperation of a Mobile Robot: A User Study. Presence: Teleoperators & Virtual Environments 14(3), 345–365 (2005)
9. Moon, I., Lee, M., Ryu, J., Mun, M.: Intelligent Robotic Wheelchair with EMG-, Gesture-, and Voice-based Interfaces. In: IEEE/RSJ Int'l. Conference on Intelligent Robots and Systems (IROS), Las Vegas, USA (2003)
10. Sato, E., Yamaguchi, T., Harashima, F.: Natural Interface Using Pointing Behavior for Human-Robot Gestural Interaction. IEEE Transactions on Industrial Electronics 54(2), 1105–1111 (2007)
11. Shiratori, T., Hodgins, J.K.: Accelerometer-based user interfaces for the control of a physically simulated character. ACM Transactions on Graphics (SIGGRAPH Asia 2008) 27(5) (2008)
12. Singh, R., Seth, B., Desai, U.: A Real-Time Framework for Vision based Human Robot Interaction. In: IEEE/RSJ Int'l. Conference on Intelligent Robots and Systems (IROS), Beijing, China (2006)
13. Song, T.R., Park, J.H., Jung, S.M., Jeon, J.W.: The Development of Interface Device for Human robot Interaction. In: Int'l. Conference on Control, Automation and Systems, Seoul, Korea (2007)
14. Stiefelhagen, R., Ekenel, H.K., Fugen, C., Gieselmann, P., Holzapfel, H., Kraft, F., Nickel, K., Voit, M., Waibel, A.: Enabling Multimodal Human Robot Interaction for the Karlsruhe Humanoid Robot. IEEE Transactions on Robotics 23(5), 840–850 (2007)
15. Varcholik, P., Barber, D., Nicholson, D.: Interactions and Training with Unmanned Systems and the Nintendo Wiimote. In: Interservice/Industry Training, Simulation, and Education Conference, I/ITSEC (2008)

A Cognitively Motivated Route-Interface for Mobile Robot Navigation

Mohammed Elmogy, Christopher Habel, and Jianwei Zhang

Abstract. A more natural interaction between humans and mobile robots can be achieved by bridging the gap between the format of spatial knowledge used by robots and the format of languages used by humans. This enables both sides to communicate by using shared knowledge. Spatial knowledge can be (re)presented in various ways to increase the interaction between humans and mobile robots. One effective way is to describe the route verbally to the robot. This method can permit computer language-naive users to instruct mobile robots, which understand spatial descriptions, to naturally perform complex tasks using succinct and intuitive commands. We present a spatial language to describe route-based navigation tasks for a mobile robot. The instructions of this spatial language are implemented to provide an intuitive interface with which novice users can easily and naturally describe a navigation task to a mobile robot in a miniature city or in any other indoor environment. In our system, the instructions of the processed route are analyzed to generate a symbolic representation via the instruction interpreter. The resulting symbolic representation is supplied to the robot motion planning stage as an initial path estimation of route description and it is also used to generate a topological map of the route's environment.

1 Introduction

A more natural interaction between humans and mobile robots – with the least collective effort – can be achieved if there is a common ground of understanding [11, 2]. A natural language interface supports more natural styles of interaction between

Mohammed Elmogy · Christopher Habel · Jianwei Zhang
Department of Informatics, University of Hamburg, Vogt-Kölln-Straße 30, D-22527 Hamburg, Germany
e-mail: {elmogy,habel,zhang}@informatik.uni-hamburg.de

H. Ritter et al. (Eds.): Human Centered Robot Systems, COSMOS 6, pp. 73–82.
springerlink.com

robots and their users. Route descriptions are considered as one of the more important natural language interfaces between humans and mobile robots for applying an effective human-robot interaction.

To describe a navigation task to a mobile robot, route instructions are used to specify the spatial information about the route environment and the temporal information about the move and turn actions which will be executed by the robot [8]. Good route instructions should contain adequate information on these two aspects by considering the spatial environment of the robot and the relevant navigation and perception actions. To express the route in an effective way, the rules and sequence of commands should be expressed very concisely. Natural language uses symbols and syntactic rules to interact with the robots which dispose of represented knowledge at the symbolic level.

On the other hand, spatial reasoning on the natural language route is essential for both humans and mobile robots. Spatial reasoning gives robots the ability to use human-like spatial language and provides the human user with an intuitive interface that is consistent with his innate spatial cognition [12]. It can also accelerate learning by using symbolic communication, which has been shown in [3].

This paper is organized as follows. Section 2 discusses some current implementations of natural language interfaces for both mobile robots and simulated artificial agents. In section 3, the structure of our route instruction language (RIL), which is used to describe the route for the mobile robot, is presented. Section 4 discusses the creation of the symbolic representation of the route. The grounding of the symbolic representation with the perceptual data in the physical environment is illustrated in section 5. Finally, the conclusion is presented in section 6.

2 Related Work

In the last three decades, there has been considerable research on spatial language and spatial reasoning. This motivates the research interest of using spatial language for interacting with artificial navigational agents. Many researchers [12, 21, 18, 17] have proposed frameworks using natural language commands in simulated or real-world environments to guide their artificial agents during navigation. In this section, some implementations of natural language interfaces for mobile robots and simulated agents will be discussed.

In our group, Tschander et al. [21] proposed the idea of a cognitive-oriented Geometric Agent (GA) which simulates instructed navigation in a virtual planar environment. This geometric agent can navigate on routes in its virtual planer environment according to natural-language instructions presented in advance. In their approach, Conceptual Route Instruction Language (CRIL) is used to represent the meaning of natural language route instructions. It combines the latter with the spatial information gained from perception to execute the desired route. Tellex and Roy [18] implemented spatial routines to control the robot in a simulator. They defined a lexicon of words in terms of spatial routines and used that lexicon to build a speech-controlled robot in a simulator. Their system is unified by a high-level module that receives the output from the speech recognition system and simulated sensor data, creates a script using the lexicon and the parse structure of the command, and then sends appropriate commands to the simulated robot to execute that command. However, their current implementation acts only on the current snapshot of sensor readings which leads to errors in the robot's behavior.

On the other hand, there are considerable research efforts in developing various command sets for mobile robots and robotic wheelchairs [15, 19, 14, 16]. The mobile robot community has created systems that can understand natural language commands. Many research efforts [21, 17, 1, 20] focus on using spatial language to control the robot's position and behavior, or to enable it to answer questions about what it senses. In general, previous work in this area has focused on developing various command sets for mobile robots and robotic wheelchairs, without directly addressing aspects of language that are context-sensitive. Torrance [20] implemented a system that is capable of mediating between an unmodified reactive mobile robot architecture and domain-restricted natural language. He introduced reactive-odometric plans (ROPs) and demonstrates their use in plan recognition. The communication component of this architecture supports a typewritten natural language discourse with people. This system was brittle due to place recognition from odometric data and the use of IR sensors for reactive motion control. The resulting ROPs do not contain error-reducing stopping conditions, and this has caused problems in some parts of the tested environment where hallways do not sufficiently constrain the reactive navigation system.

Skubic et al. [17] implemented robot spatial relationships combined with a multimodal robot interface that provides the context for the human-robot dialog. They showed how linguistic spatial descriptions and other spatial information can be extracted from an evidence grid map and how this information can be used in a natural human-robot dialog. With this spatial information and linguistic descriptions, they established a dialog of spatial language. To overcome the object recognition problem (the system does not support vision-based object recognition), they have defined a class of persistent objects that are recognized and named by the user.

3 Route Instruction Language (RIL)

In our system, we present a spatial language – called Route Instruction Language (RIL) [7] – to describe route-based navigation tasks for a mobile robot. This language is implemented to present an intuitive interface that will enable novice users to easily and naturally describe a route to a mobile robot in indoor and miniature city environments. We proposed this language to avoid ambiguity and misunderstanding during route description. Therefore, a non-expert user can describe the route for the mobile robot by using simple and easy to understand instructions. Fig. 1 shows an overview structure of our system.

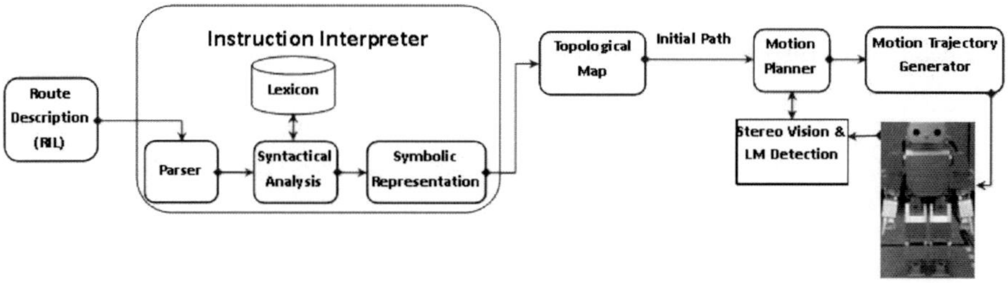

Fig. 1 System architecture

The RIL is developed to describe the route between the start and end points to a mobile robot. It is intended as a semi-formal language for instructing robots, to be used via a structured graphical user interface. RIL provides elementary instruction statements which are processed to supply the robot with a sequence of motion actions. During navigation, this sequence of actions is processed by the motion planner to determine the footstep placements which will be effected by the humanoid robot to execute the route. Each statement in the RIL constitutes a spatial instruction which relates verbally coded motion concepts to one or more landmarks by use of a suitable spatial relationship.

Table 1 RIL command set and their syntax

Command Type	Command Name	Syntax
Position	$START()	$START ([Pre1|Direction], Landmark1, [Pre2], [Landmark2])
	$STOP()	$STOP (Pre1|Direction, Landmark1, [Pre2], [Landmark2])
	$BE()	$BE (Pre1|Direction, Landmark1, [Pre2], [Landmark2])
Locomotion	$GO()	$GO([Count], [Direction]| [Pre1], [Landmark1], [Pre2],[Landmark2])
	$CROSS()	$CROSS ([Pre1], Landmark1, [Pre2], [Landmark2])
	$PASS()	$PASS ([Pre1], Landmark, direction, [Pre2], [Landmarket2])
	$FOLLOW()	$FOLLOW ([Landmark1], Pre, Landmark2)
Change of Orientation	$ROTATE()	$ROTATE (Direction, Pre, Landmark)
	$TURN()	$TURN ([Count], [Pre1], Direction, [Pre2], [Landmark])

The commands of the RIL and their syntax are shown in Table 1. Each instruction of the RIL specifies motion verbs, directions, destinations, and landmarks. The RIL commands are divided into three basic types: position, locomotion, and change of orientation commands. The position commands are used to indicate the current position of the robot during navigation. They are also used to determine the start and end points of the route. The Locomotion commands are used to instruct the robot to move in the spatial environment in a specific direction or to follow a certain path. The last category is the change of orientation commands, which are used to rotate around a landmark or turn in a certain direction.

The command syntax consists of a command word and an arbitrary number of arguments as shown in Table 1. The command word indicates the action which will be taken by the mobile robot and is represented in the imperative form of the verb, e.g., GO, TURN, BE, etc. Each argument is a place holder for a specific group of words such as prepositions, directions, the number of turns, and landmarks. To add more flexibility to the command syntax, multiple kinds of command syntax have been defined. Mandatory arguments are typed without any brackets, whereas optional arguments are placed between rectangular brackets '[]'. The pipe symbol '|' indicates an OR operator. Fig. 2 shows an example of a route description from the railway station to the McDonald's restaurant in our miniature city using RIL.

(a) Miniature City Map

```
$START( RailwayStation, right)
$GO( forward, into, Street)
$PASS( Saturn, left)
$PASS( BurgerKing, right)
$PASS( C&A, left)
$BE( at, CrossRoads)
$GO( forward, into, Street)
$PASS( Church, left)
$PASS( KarStadt, right)
$BE( at, CrossRoads)
$TURN( right)
$GO( forward, into, Street)
$PASS( TownHall, left)
$STOP( McDonalds, left)
```

(b) Route Description

Fig. 2 A route description from the railway station to the McDonald's restaurant in our miniature city using RIL

We carried out an experiment to test the suitability of RIL for communicating a route description to a robot. 18 participants took part in the experiment (age 22 to 35 years). None of the participants had any background knowledge on route instructions and robotics. First, we gave them a description of the RIL syntax, a map of the miniature city, and an example of a suitable route description. We asked them to describe a route between the railway station and the McDonald's restaurant in the miniature city as depicted in Fig. 2(a). 89% of the participants described the route correctly, but the rest are confused about how to use some commands and parameters. 83% of the participants stated that the RIL is simple and easy to learn, but the rest of them preferred to use a controlled natural language without any specific syntax for the instructions. 78% of the participants agreed that it is better to provide the commands of RIL with many optional parameters than to restrict them to a single syntax.

4 Instruction Interpreter

The instruction interpreter is used to discriminate, identify, and categorize the motion actions of the processed route description. It combines definitions from the lexicon according to the parse structure of the instruction, creating a symbolic script that describes the navigation process. The generated symbolic representation is used to create a topological map for the route environment. It is also supplied to the motion planner as an initial path estimation of the navigation task to help in generating the footstep placements for the humanoid robot. This symbolic script is based on CRIL representation which was developed by our group [21].

The instruction interpreter contains a simple parser, a lexical analysis, and a syntactic analysis. The parser is supplied by the route description text. It separates the text into individual instructions. Each instruction is split into sequence of words using space and punctuation characters as delimiters. The resulting list is entered at the syntactical analysis stage to identify the structure of instructions by comparing their structure with a list of all kinds of instruction syntax which are understandable by the robot. Each word is looked up in the lexicon to obtain its type and features. The available types of words in the lexicon are command verbs, directions, prepositions, numbers of turns, and landmarks. Each verb entry in the lexicon consists of an action verb and an associated script composed from the set of its primitives and depends on the specified arguments passed to its instruction. It is defined as a script of primitive operations that run on data extracted from the analyzed instruction.

After analyzing the route instructions syntactically and connecting each resulting verb with its motion procedure, the symbolic representation of the route is generated. The resulting symbolic script consists of three basic components: motion actions, spatial relationships, and landmarks. The motion actions are classified into the following four different actions:

- BE_AT Action: It presents the position of the robot during navigation. It identifies the start, current, and end positions of the robot during navigation.
- GO Action: It indicates the motion actions which should be taken by the mobile robot.
- VIEW Action: It is used to notice a landmark in a certain direction or region during navigation.
- CH_ORIENT Action: It is used to indicate a change in the current orientation of the mobile robot motion during navigation based on a specific direction or landmark.

The spatial relationships are classified into two types. First, relations represent a location with respect to a landmark. Second, relations specify a direction with respect to one or two landmarks. Finally, the landmark features are retrieved from the knowledge base. They contain data about their shape, color or color histogram, and recognition method values. In addition to the retrieved features, the relationship feature is extracted from the processed route to describe the relation between the current processed landmark and other landmarks in the same path segment. It is used to handle uncertainty and missing information during the robot navigation.

Landmarks in our miniature city are classified into definite and indefinite landmarks depending on their features. Definite landmarks have unique characteristics which single them out from among the other landmarks in the miniature city, such as the Burger king restaurant, the Saturn store, and the town hall. On the other hand, indefinite landmarks have a number of properties that are not unique such as buildings, crossroads, and streets.

After creating the symbolic representation of the route, the robot requires an adequate representation of the route environment. This representation should be abstract enough to facilitate higher-level reasoning tasks like strategic planning or situation assessment, and still be detailed enough to allow the robot to perform lower-level tasks like path planning/navigation or self-localization [13]. The topological map representation is used to describe relationships among features of the environment in a more abstract representation without any absolute reference system [22]. Our implementation of the topological map represents the robot's workspace in a qualitative description. It presents a graph-like description of the route where nodes correspond to significant, easy-to-distinguish landmarks, and arrows correspond to actions or action sequences which will be executed by the mobile robot [6].

5 Symbol Grounding

After building the topological map, the resulting symbolic representation is supplied to the motion planner as initial path estimation. The motion planner uses both the symbolic representation and the output of the stereo vision and landmark recognition stage to calculate the desired footstep placements of the humanoid robot to execute the processed route.The motion planner grounds the landmark symbols to their corresponding physical objects in the environment. Therefore, the symbolic and physical presentations of the landmarks should be integrated. Many researchers [9, 4, 10] have worked on the symbol grounding problem to solve the problem of incorporating the high-level cognitive processes with sensory-motoric processes in robotics. Cognitive processes perform abstract reasoning and generate plans for actions. They typically use symbols to denote objects. On the other hand, sensory-motoric processes typically operate from sensor data that originate from observing these objects. The researchers tried to maintain coherence between representations that reflect actions and events, and the produced stream of sensory information from the environment. Accordingly, mobile robots need learning abilities that constrain abstract reasoning in relation to dynamically changing external events and the results of their own actions [12].

Harnard [9] considered perceptual anchoring as an important special case of symbol grounding. The anchoring is defined as the process of creating and maintaining the correspondence between symbols and sensor data that refer to the same physical objects [5]. We used a perceptual anchor to incorporate the symbols of the landmarks represented in the symbol system (Σ) and the physical landmarks retrieved from the perceptual system (Π). The predicate grounding relation (g) is used to encode the correspondence between predicate symbols and admissible values of observable attributes. It contains the values of the landmark properties, such as color histogram values, shape, area range, and recognition method values. We used a color histogram, scale invariant features transform (SIFT), and the bag of features methods to recognize the landmarks. As shown in Fig. 3, the perceptual anchoring (α) for a landmark contains a pointer to its symbol (δ), a pointer to its physical object (π), and its signature (γ).

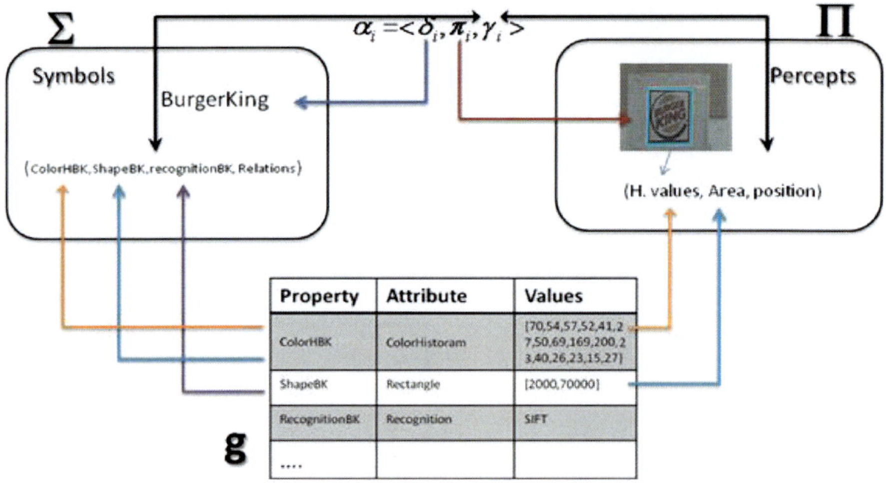

Fig. 3 Anchoring process between the symbolic and perceptual systems

6 Conclusion

We presented RIL, a semi-formal language to be used by non-expert users to instruct mobile robots. Based on RIL, we designed and realized an intuitive interface to mobile robots preventing misunderstanding and ambiguities in route descriptions. Starting from a set of commands, the instruction interpreter stage performs the analysis of route instructions and its lexicon relates the internal procedures to perceptual objects and specifies actions that can be carried out by the mobile robot. The instruction interpreter analyzes the route to generate its equivalent symbolic representation which is supplied to the motion planner as initial path estimation.

The resulting symbolic representation of the route is used to generate a graphical representation of the route to supply the robot with global route information and to prevent it from getting trapped in local loops or dead-ends in unknown environments. Finally, the symbolic representation is supplied to the motion planner to ground the landmark symbols to their equivalent physical objects by using perceptual anchoring.

References

1. Bischoff, R., Jain, T.: Natural communication and interaction with humanoid robots. In: Second International Symposium on Humanoid Robots, Tokyo (1999)
2. Brennan, S.E.: The grounding problem in conversations with and through computers. In: Fussell, S.R., Kreuz, R.J. (eds.) Social and cognitive psychological approaches to interpersonal communication, pp. 201–225 (1991)
3. Cangelosi, A., Harnad, S.: The adaptive advantage of symbolic theft over sensorimotor toil: grounding language in perceptual categories. Evolution of Communication 4, 117–142 (2000)
4. Chella, A., Coradeschi, S., Frixione, M., Saffiotti, A.: Perceptual anchoring via conceptual spaces. In: Proceedings of the AAAI 2004 workshop on anchoring symbols to sensor data (2004)
5. Coradeschi, S., Saffiotti, A.: An introduction to the anchoring problem. Robotics and Autonomous Systems 43, 85–96 (2003)
6. Elmogy, M., Habel, C., Zhang, J.: Robot topological map generation from formal route instructions. In: Proceedings of the 6th international cognitive robotics workshop at 18th european conference on artificial intelligence (ECAI), Patras, Greece, pp. 60–67 (2008)
7. Elmogy, M., Habel, C., Zhang, J.: Spatial Language for Route-Based Humanoid Robot Navigation. In: Proceedings of the 4th international conference on spatial cognition (ICSC 2009), Roma, Italy (to be published)
8. Habel, C.: Incremental Generation of Multimodal Route Instructions. In: Natural language Generation in Spoken and Written dialogue, AAAI Spring Symposium 2003, Palo alto, CA, pp. 44–51 (2003)
9. Harnad, S.: The symbol grounding problem. Physica D. Nonlinear phenomena 42(1-3), 335–346 (1990)
10. Karlsson, L., Bouguerra, A., Broxvall, M., Coradeschi, S., Saffiotti, A.: To secure an anchor - a recovery planning approach to ambiguity in perceptual anchoring. AI Communications 21(1), 1–14 (2008)
11. Kiesler, S.: Fostering Common Ground in Human-robot Interaction. In: Proceedings of the IEEE International Workshop on Robot and Human Interactive Communication (ROMAN), pp. 729–734 (2005)
12. Lauria, S., Bugmann, G., Kyriacou, T., Bos, J., Klein, E.: Training Personal Robots using Natural Language Instruction. IEEE Intelligent Systems 16, 38–45 (2001)
13. MacMahon, M.: Marco: A Modular Architecture for following Route Instructions. In: Proceedings of the AAAI Workshop on Modular Construction of Human-like Intelligence, Pittsburgh, PA, pp. 48–55 (2005)
14. Pires, G., Nunes, U.: A Wheelchair Steered Through Voice Commands and Assisted by a Reactive Fuzzy-logic Controller. Journal of Intelligent and Robotic Systems 34, 301–314 (2002)
15. Schulz, R., Stockwell, P., Wakabayashi, M., Wiles, J.: Towards a Spatial Language for Mobile Robots. In: Proceedings of the 6th international conference on the evolution of language, pp. 291–298 (2006)
16. Simpson, R.C., Levine, S.P.: Adaptive shared control of a smart wheelchair operated by voice control. In: Proceedings of the 1997 IEEE/RSJ international conference on intelligent robots and systems (IROS 1997), pp. 622–626 (1997)
17. Skubic, M., Perzanowski, D., Blisard, S., Schultz, A., Adams, W., Bugajska, M., Brock, D.: Spatial Language for Human-robot Dialogs. IEEE Transactions on Systems, Man, and Cybernetics, Part C: Applications and Reviews 34(2), 154–167 (2004)

18. Tellex, S., Roy, D.: Spatial Routines for a Simulated Speech-Controlled Vehicle. In: Proceedings of the 1st ACM sigchi/sigart conference on human-robot interaction, Salt Lake City, Utah, USA, pp. 156–163 (2006)
19. Tellex, S., Roy, D.: Grounding Language in Spatial Routines. In: Proceedings of AAAI Spring Symp. on Control Mechanisms for Spatial Knowledge Processing in Cognitive / Intelligent Systems (2007)
20. Torrance, M.C.: Natural communication with mobile robots. MIT Department of Electrical Engineering and Computer Science (1994)
21. Tschander, L.B., Schmidtke, H., Habel, C., Eschenbach, C., Kulik, L.: A geometric agent following route instructions. In: Freksa, C., Brauer, W., Habel, C., Wender, K.F. (eds.) Spatial Cognition III. LNCS (LNAI), vol. 2685, pp. 89–111. Springer, Heidelberg (2003)
22. Zavlangas, P.G., Tzafestas, S.G.: Integration of topological and metric maps for indoor mobile robot path planning and navigation. In: Vlahavas, I.P., Spyropoulos, C.D. (eds.) SETN 2002. LNCS (LNAI), vol. 2308, pp. 121–130. Springer, Heidelberg (2002)

With a Flick of the Eye: Assessing Gaze-Controlled Human-Computer Interaction

Hendrik Koesling, Martin Zoellner, Lorenz Sichelschmidt, and Helge Ritter

Abstract. Gaze-controlled user interfaces appear to be a viable alternative to manual mouse control in human-computer interaction. Eye movements, however, often occur involuntarily and fixations do not necessarily indicate an intention to interact with a particular element of a visual display. To address this so-called Midas-touch problem, we investigated two methods of object/action selection using volitional eye movements, fixating versus blinking, and evaluated error rates, response times, response accuracy and user satisfaction in a text-typing task. Results show significantly less errors for the blinking method while task completion times do only vary between methods when practice is allowed. In that case, the fixation method is quicker than the blinking method. Also, participants rate the fixation method higher for its ease of use and regard it as less tiring. In general, blinking appears more suited for sparse and non-continuous input (e.g., when operating ticket vending machines), whereas fixating seems preferable for tasks requiring more rapid and continuous selections (e.g., when using virtual keyboards). We could demonstrate that the quality of the selection method does not rely on efficiency measures (e.g., error rate or task completion time) alone: user satisfaction measures must certainly be taken into account as well to ensure user-friendly interfaces and, furthermore, gaze-controlled interaction methods must be adapted to specific applications.

1 Introduction

Eye tracking or *oculography* is a common method that monitors and records movements of the eye ball in humans. From this raw data, *gaze positions* on objects or

Hendrik Koesling · Martin Zoellner · Helge Ritter
Faculty of Technology, Bielefeld University, Germany
e-mail: {ihkoesli,mzoellne,helge}@techfak.uni-bielefeld.de

Lorenz Sichelschmidt
Faculty of Linguistics and Literary Studies, Bielefeld University, Germany
e-mail: max.sichelschmidt@uni-bielefeld.de

H. Ritter et al. (Eds.): Human Centered Robot Systems, COSMOS 6, pp. 83–92.
springerlink.com

on a display screen that an observer currently looks at can easily be computed (e.g., Stampe, 1993). Most *eye-tracking systems* that are available now provide the required geometric transformations and can compute gaze positions on-line in real time. Data is processed with a high temporal resolution (up to 4 kHz, normally between 50 Hz and 500 Hz), a spatial resolution of typically around 0.05 degrees of visual angle and a spatial accuracy of measurements of below 0.8 degrees. Based on the eye-mind hypothesis proposed by Just and Carpenter (1980), stating that eye fixations on a scene are informative as to what content is currently being processed, eye tracking is now a well established method in basic research. It is applied in a wide field of scientific studies concerned with, for example, reading (Rayner, 1997), scene perception (Underwood, 2005), visual search (Pomplun et al., 2006) or attention modeling (e.g. Itti, Koch & Niebur, 1998). Furthermore, eye tracking is a common method in medicine for diagnostics and treatment (e.g., Korn, 2004) and has also found its way into consumer research and marketing (e.g., Rettie & Brewer, 2000).

With its high-precision measurements and the on-line availability of gaze-position data, eye tracking is also a suitable interface in human-computer interaction (e.g., Zhai, 2003). Eye movements are already used in a number of applications to control technical devices. Most common is the use of gaze-controlled (or *gaze-contingent*) devices by people with special needs, in particular patients with restricted (manual) interaction capabilities. A gaze-controlled interface presents a suitable means for them to communicate with others (via computer) or to control functions of assistance systems that allow them to live independently from constant caretaking.

The present study addresses the questions how we can improve gaze-controlled interaction with computers or technical devices - and, if these devices are used as means for communication with others, how to improve human-human interaction in *mediated communication* scenarios. Gaze control on a computer display screen normally implies to engage in interaction with a particular screen area (a specific window or button, for example) by directing one's gaze (normally associated with the focus of *attention*) to it. The current gaze position on the screen thus represents the computer mouse cursor position, moving a user's gaze about mimicking mouse movements. This method is generally referred to as the "gaze mouse".

One of the major problems in gaze-controlled computer interaction is how to conveniently and reliably execute events and actions, that is, how to execute mouse button clicks. The gaze position only provides information about the screen area that a user currently looks at or attends to. However, it does not necessarily indicate whether the user wants to interact with that particular area. This difficulty to be able to unambiguously attribute a particular intention to a gaze point, for example initiating an action such as selecting a letter from a virtual keyboard, is known as the *"Midas-touch problem"*. Without further processing or analysis of the eye-tracking data it is not possible to identify the user's intent. In order to detect this intent, we must be able to distinguish between the user's pure focus of attention and the user's volition to execute a certain action (Jacob, 1995).

Several of such *volitional action-selection methods* have been implemented in the past (c.f. Vertegaal, 2002). One of the most common is referred to as the "fixation method". This method makes use of the fact that, at least when observing static images, the gaze moves about in a particular pattern of alternating *fixations* and *saccades*. During fixations, the gaze remains stationary at a specific location for between 80 ms and 250 ms. A fixation is normally followed by a saccade, a rapid, ballistic movement of the eye to the next fixation point. Saccades usually last around 30 ms, depending on the saccade amplitude. Healthy humans only perceive visual information during fixations but not during saccades – "saccadic suppression" (Burr, 2004). This typical pattern of saccadic eye movements is taken advantage of in the fixation method. As long as the user's gaze roams about the stimulus display without stopping at fixation points for more than, let us say 500 ms, no action in that particular area is being executed. When a user wants to, for example, activate a button, he or she has to simply hold a fixation for longer than the "activation" threshold (here 500 ms). This gives the user sufficient time to attend to a particular area on the screen, perceive all relevant information at that particular location and, when no action is intended, to saccade to another area before any action is taken. Another common method is the "blinking method". Rather than selecting actions by prolonged fixations it uses volitional eye blinks.

Other methods also exist, often variants of the fixation or blinking method. A convenient one uses gaze control for mouse-cursor movements only. Action selection then relies on manual input such as a mouse-button press. Another gaze-controlled action-selection method requires the definition of a separate "action region" located next to the appropriate interaction area. User have to specifically direct their gaze to the action region in order to initiate an action, thus avoiding erroneous triggering of actions by simply looking at an interesting spot. However, this method can only be used when the display is sparsely "populated" with interaction areas and provides sufficient space between interaction areas. For virtual keyboards in gaze-typing applications, the use of this method is limited (e.g., Helmert et al., 2008).

The previous paragraphs made clear that one of the problems with most eye movements and blinks is that they often occur involuntarily, so that gaze-controlled interaction can be error-prone and often initiate unintended actions. Activation thresholds must thus be adjusted carefully. Furthermore, and partly due to conservative threshold settings, gaze-contingent interaction is often slow. This is no problem when only few actions are required such as interacting with a vending machine at a train station. If, however, frequent, rapid actions are necessary, for example, in character selection for gaze-typing, this drastically slows down performance. To overcome this restriction, adaptive methods exist that detect typical fixation or blink times of each individual user and reduce activation thresholds to an optimal time that still allows the user to perceive all relevant information. Some of the presented methods are being used in commercially available gaze-controlled user interfaces. However, most methods have not been extensively evaluated with regard to their ease of use, efficiency or user satisfaction. Furthermore, no guidelines have yet been proposed as to when which method should be preferred over another one.

The present study investigated these aspects and compared the blinking and fixation methods in a typical computer-based interaction task: text (gaze) typing. This required to develop a screen-based, virtual keyboard user interface and integrate it into the experimental programming environment of the eye tracker. Based on this experimental environment we specifically addressed the research questions which of the two methods is preferable, taking into account its efficiency, how easy it is to learn and its user-friendliness/satisfaction. In the experimental study, participants had to gaze-type sentences using either of the two fixation and blinking methods. Subsequently, they filled in a questionnaire that focussed on querying usability aspects of the applied gaze-contingent interaction methods.

2 Method

Participants: 20 native German speakers participated in the experiment, 10 females and 10 males aged between 22 and 35 years (average 24.3 years). All participants had normal or corrected-to-normal vision. The participants were naive to the task.

Apparatus: We used a remote LC Technologies EyeGaze eye-tracking system to monitor the participants' eye movements during the experiment. Eye movements were sampled binocularly at 120 Hz (interlaced). Before the start of the experiment and before each trial, a multi-point calibration procedure was performed to enable accurate data recordings. Stimuli were shown on a 17-inch Samsung SyncMaster 731BF TFT-screen. The screen resolution was set to 1024 x 768 pixels at a refresh rate of 60 Hz. Subjects were seated approximately 60 cm from the screen. The screen subtended a visual angle of 31.6 degrees horizontally and 23.1 degrees vertically. In order to minimise head movements and improve the accuracy of the gaze-point measurements, the participants' head was stabilised using a chin rest during the experiment. Figure 1 (left) shows the experimental setting.

Stimuli and Design: The stimuli depicted a gaze-controlled text-typing interface and consisted of three separate areas. The upper quarter of each stimulus screen

Fig. 1 Left: Experimental setting during the gaze-typing task. A participant is seated in front of the screen that displays the gaze-typing interface. Two miniature cameras of the LC Technologies eye tracker are mounted below the screen and monitor the user's eye movements. Right: Stimulus display screen during a trial. The top frame marks the task area, the middle frame the input control area and the bottom frame the interaction area. Frames are inserted for illustration purposes only and were not visible during the experiment. Some text has already been typed and appears in the input control area. The currently selected character 'I' is marked by a green surround (dotted here for illustration) to provide feedback to the user about the successful selection.

constituted the "task area" (width 21 degrees, height 5 degrees). It contained a German target sentence that was constantly present throughout each trial. Each target sentence consisted of 6 to 11 words and contained 55 to 65 characters, all in upper case. Character size measured approximately 0.5 degrees in height and width. Sentences were taken from local newspapers and were syntactically and grammatically correct. A different sentence was used in each trial, sentences also varied between the fixation and blinking method conditions. Sentence lengths – as measured by the number of characters – in the both conditions were matched. All participants viewed the same target sentences, however, presentation order was randomised.

An "input control area" was placed below the task area and had the same size as the task area (21 x 5 degrees). At the beginning of each trial this area was blank. It filled with characters when participants selected keys from the "interaction area". The interaction area covered the lower half of the screen (21 x 10 degrees) and contained a virtual keyboard with a German "QWERTZ"-style layout. The keyboard only contained upper case characters, the space bar and a sentence terminator key ('.'), but no shift, backspace or other keys. Character keys had a size of 2 degrees in height and width and were separated by a gap measuring 1 degree. The space bar measured 2 degrees in height and 13.5 degrees in width. Participants could select keys from the interaction area by the gaze-contingent selection methods (fixation or blinking). Only the interaction area "responded" to gaze events which were prompted in the input control area. Figure 1 (right) shows a typical display screen.

Procedure: Prior to the experiment, the chin rest and the eye-tracking system were adjusted so as to guarantee a comfortable viewing position for participants and accurate eye-movement measurements. Participants were then instructed about their task: They had to type a target sentence that was presented in the task area on the screen using either the fixation or the blinking gaze-typing method. Participants should accomplish this task as accurately and quickly as possible.

All participants completed two blocks of trials, one block using the fixation method for gaze-typing, the other block using the blinking method. The sequence of blocks was counter-balanced so that 5 participants of each gender group started with the fixation method, the other 5 participants started with the blinking method. Each block consisted of a calibration procedure, followed by a single practice trial and 5 experimental trials. The calibration established a correspondence between eye-ball movements and gaze points on the screen for the subsequent measurements.

The practice and experimental trials started with a fixation cross at the centre of a blank screen for 500 ms. The stimulus screen then appeared and showed the target sentence in the task area. Participants then started to select the character keys from the interaction area using the appropriate method (fixation or blinking) for the respective experimental block. The successful gaze-selection of a character from the interaction area was signaled by a green frame appearing around the selected character key for 300 ms and was also prompted by the respective character appearing in the input control area. Participants iterated the selection process until they had completely re-typed the target sentence and pressed a manual response key (mouse button). Corrections of typing errors during gaze-typing were not possible. No gaze cursor was shown to provide feedback about the current gaze position.

In a pre-experiment, we found that the average duration of involuntary blinks as they normally occur for retaining the liquid film on the cornea or for cleaning purposes lies below 70 ms. This is also well below fixation durations when viewing static scenes (80–250 ms, also see Section 1). This led us to setting the "activation" threshold for character-key selection to 300 ms for both the fixation and blinking methods. A character key was thus only then selected for gaze-typing when either a fixation or a blink occurred that lasted more than 300 ms. This threshold setting would ensure that only volitional blinks and fixations (i.e., those that were intended to select an action) could activate character keys. Defining the same activation threshold for both methods ensured their easy comparison. As obviously no gaze point coordinates are present during blinks, the last valid coordinates before a blink were taken to indicate where a blink occurred.

After the experiment, participants completed a questionnaire. Questions focused on user-satisfaction and ease-of-use of each of the two individual methods and how the two methods compared. In total, it took participants between 30 and 40 minutes to complete the experiment and the questionnaire.

Data analysis: Statistical data analyses were computed using SPSS 14.0. We used t-tests for paired samples to compare means between the *gaze-typing methods* (within-subjects factor, fixation vs. blink). We also compared means between subjects for the *gender groups* (male vs. female) and for the *order of methods* (fixation trials before blink trials vs. blink trials before fixation trials). We analysed error rates (ER, number of errors per sentence), completion time (CT, in seconds) and the categorised responses from the questionnaire (for factor gaze-typing method only) as dependent variables. The α-level for all t-tests was set to 0.05.

3 Results and Discussion

Accumulated over all gender groups and method orders, participants made 5.20 typing errors per sentence using the fixation method. For the blinking method, ER reached 2.86 on average, only slightly more than half the ER for the fixation method (see Figure 2, left). The comparison of means using the t-test confirmed that ERs were significantly different between methods ($t(19) = 4.187; p < 0.001$).

A closer inspection of data revealed that for the group that started with the blinking method, average ER for the blinking method reached 3.23 and 5.14 for the fixation method. These means were not significantly different, however, still showed a strong tendency ($t(9) = 2.124; p < 0.063$). In contrast, the group that started with the fixation method recorded average ERs for the blinking method of 2.45 and 5.27 for the fixation method. These means were significantly different from each other ($t(9) = 4.351; p < 0.001$).

No significant differences existed between female and male participants. Within both groups the blinking method produced significantly less errors than the fixation method. Within female group: ($t(9) = 2.480; p < 0.038$). Within male group: ($t(9) = 3.297; p < 0.001$).

Inspecting data during the course of the experiment, data seems to suggest practice and fatigue effects on ER (see Figure 2, centre). More specifically, we did notice a reduction in error rates ER towards the middle of the experiment (improvement through practice), followed by an increase in ER towards the end of the experiment (possibly caused by fatigue), reaching similar ER levels as observed at the beginning. However, in order to validate the persistence of these effects, further experiments with more trials should be conducted.

These results indicated that participants produced fewer errors in gaze-typing when they could use the blinking method instead of the fixation method. Using the blinking method rather than the fixation method first, however, seemed to compensate the disadvantage of the fixation method relative to the blinking method. Furthermore, when the fixation method was used first, absolute values of ER increased compared to ERs for the reversed order of methods. We can therefore speculate that the higher error rates in the fixation method may negatively effect the blinking method. Thus, if users indeed use (or learn) both gaze-typing methods, it appears to be convenient to start with the blinking method so that the fixation method may benefit from it. Taken into account questionnaire response data (*"Which of the two methods you felt was the less error-prone?"*), the subjective assessment of participants confirmed that the blinking method must be preferred over the fixation method (by 79% of all participants) with regard to achieving lower error rates.

The comparison of sentence completion times CT (time to gaze-type a sentence) did not show statistically significant differences between the fixation and blinking methods. Participants took, on average, 112 seconds to type a single sentence with the fixation method and 123 seconds with the blinking method (see Figure 2, right).

Interestingly, separating the data by the order of methods led to a significant difference between the fixation and blinking methods – however, only when the blinking method was used first. In that case, the fixation method (93 s) was significantly quicker ($t(9) = 4.507; p < 0.001$) than the blinking method (121 s). For the reversed order of methods (fixation method before blinking method), no significant difference existed. In fact, in that case absolute CTs were rather long for both methods: 132 s for the fixation method and 127 s for the blinking method.

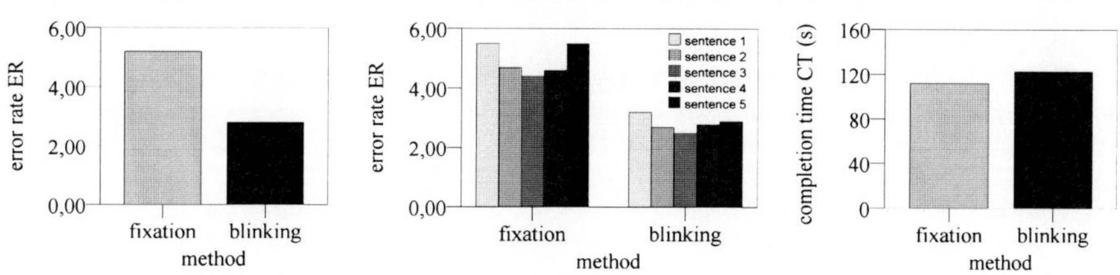

Fig. 2 Left: Error rate ER as a function of selection method, averaged over gender group and method order. Centre: Error rate ER as a function of stimulus sequence for fixation method (left) and blinking method (right), averaged over gender group and method order. Right: Sentence completion time CT as a function of selection method, averaged over gender group and method order.

Whereas within both female and male groups no significant differences could be established between the fixation method (females: 110 s, males: 115 s) and the blinking method (females: 143 s, males: 104 s), female participants (127 s), on average, took significantly longer ($t(9) = 2.232; p < 0.031$) to accomplish the gaze-typing task than males (109 s). This may be explicable when we take into account responses from the questionnaire. Here, we learnt that females rate their experience with computers, with typing and the time spent at a computer per week (in particular for typing tasks) lower than males do. Longer CTs in women might thus be attributed to a lack of experience with the specific keyboard-type input device.

Results from the analyses of sentence completion times were in line with the findings from the error rates: When the blinking method was used before the fixation method, the fixation method seemed to benefit from the previous (blink) trials by significantly reduced CTs, both compared to CTs for the blink method for that order of methods and to CTs for the fixation method when the order of methods was reversed. This could indicate that, after a certain practice period with gaze-contingent input interfaces, the fixation method may be preferred over the blinking method with regard to task completion speed, at least for gaze-typing. Practice with eye guidance seems to be a prerequisite to efficiently apply the fixation method. The "practice method" may indeed not have to be the fixation method itself. The pure gaze-contingent nature of such a practice method (as it is the case with the blinking method that apparently served as the practice method here) may be sufficient to improve the performance of the fixation method. A possible reason for why more practice is needed for the fixation method than for the blinking method may be that humans are not so familiar with their own fixation behaviour. Most are probably not even aware of that they fixate objects when they look at them. Thus, the concept of holding a stable view for an extended period must first be understood – and practiced – before it can properly be applied. In contrast, most humans are aware of that they blink and can easily control blinks and blink durations without practice.

Finally, the analysis of the questionnaire data revealed, for example, that only 35% of participants would consider using gaze-controlled interfaces instead of a conventional computer mouse for human-computer interaction. This is probably due to the fact that all participants were healthy individuals with no problems in manual control. They could all accomplish the same typing task using a manual input device such as a computer keyboard or mouse in a fraction of the time and with a lower error rate than when using the fixation or blinking method. For participants with motor-control deficits, however, the acceptance of the gaze-contingent interface would most certainly have been much higher.

More participants (65%) rated the fixation method as being more intuitive and easier to get used to (55%) than the blinking method - the latter rating, in fact, contradicting the apparent need to practice the fixation method as discussed above. A majority of participants also rated the fixation method as less tiring (70%). This, again, is not reflected in the empirical data - at least when considering increasing error rates towards the end of the fixation method block as an indicator for a fatigue effect. On the other hand, participants found that the blinking method had advantages over the fixation method in that it was more accurate in making character key selections (70%), less error-prone (65%) and that the blinking procedure in general caused less problems (55%) than the fixation method. Finally, participants rated the general "fluency" of the fixation method higher than that of the blinking method: 4.5 vs. 4.1 on a scale from 1 ("excellent") to 6 ("poor"). As already discussed, the participants' judgments regarding the usability of the gaze-contingent interaction methods overall well reflected the empirical data of the experiment and did not indicate any major discrepancies between subjective impressions and quantitative measures.

4 General Discussion

The analysis of the empirical data collected in the experiment demonstrated that both of the two gaze-contingent selection methods, the fixation and the blinking method, have their advantages and inconveniences in human-computer interaction. It would be difficult to prefer one over the other unless taking into account the particular demands of a specific task.

We could identify the blinking method as the one that produces fewer errors and allows for more accurate selections while, at the same time, requires less practice. If, however, time to practice is available, the fixations method may not necessarily be more error-prone. In fact, after practice, not only error rates of the fixation and blinking methods no longer significantly differ. In addition, task completion times for the fixation method also benefit from practice, they are then significantly shorter than for the blinking method. Thus, after practice, the fixation method may be more efficient than the blinking method. This, however, needs to be validated in further experiments that increase the practice time allowed. Findings so far strongly hint at the practice effect, but cannot provide unambiguous support. Undisputedly, however, the fixation method must be preferred to the blinking method. It is rated as less tiring (although this is not in line with the error rate recordings during the experiment) and more intuitive – important factors with regard to user-friendliness, user-satisfaction and possible long-term effects of novel interaction methods on users.

In conclusion, findings from the present study indicate that both gaze-contingent selection methods present feasible approaches to solve the Midas-touch problem. The blinking method appears to be well suited for applications where only few selections are needed, that does not rely on a "fluent" input stream and does not allow much time for practice. This is the case, for example, for applications such as ticket vending machines at train stations or when filling in forms that come in multiple-choice style and require ticking boxes rather than long, continuous textual input. On the other hand, the fixation method is better suited for continuous, long-lasting and rapid interaction with an application. When users are appropriately trained in using the fixation method, they may consider this interaction method an intuitive, user-friendly means of communication with their environment. Virtual keyboards, for example, in particular for users with motor control deficits, thus present a suitable application for the fixation method.

Based on the current experimental setting and the promising results with respect to the projected uses of the gaze-contingent interaction methods, we should now conduct further studies. These could, for example, review or investigate effects of additional auditory feedback about gaze cursor location, proximity to possible targets or selection success on the performance and efficiency of the interface. Other problems such as the distinction between single and double clicks – or how to implement them conveniently – must also be addressed. The present study has certainly shown that the use of gaze-contingent user interfaces in human-computer interaction is a promising method and may indeed be well suited to enhance the accessibility of technology for healthy and handicapped individuals alike. It may help handicapped people in particular to more easily communicate with their environment and thus considerably improve their quality of life.

Acknowledgements. This research was funded by the German Science Foundation (DFG Sonderforschungsbereich 673, Project A5).

References

1. Burr, D.: Eye Movements: Keeping Vision Stable. Current Biology 14, 195–197 (2004)
2. Helmert, J.R., Pannasch, S., Velichkovsky, B.M.: Influences of dwell time and cursor control on the performance in gaze driven typing. Journal of Eye Movement Research 2(4, 3), 1–8 (2008)
3. Itti, L., Koch, C., Niebur, E.: A Model of Saliency-Based Visual Attention for Rapid Scene Analysis. IEEE Trans. on Pattern Analysis and Machine Intelligence 20, 1254–1259 (1998)
4. Jacob, R.J.K.: Eye Tracking in Advanced Interace Design. In: Barfield, W., Furness III, T.A. (eds.) Virtual Environments and Advanced Interface Design, pp. 258–288. Oxford University Press, New York (1995)
5. Just, M.A., Carpenter, P.A.: A theory of reading: From eye fixations to comprehension. Psychological Review 87, 329–354 (1980)
6. Korn, H.: Schizophrenia and eye movement - a new diagnostic and therapeutic concept. Medical Hypotheses 62, 29–34 (2004)
7. Pomplun, M., Carbone, E., Koesling, H., Sichelschmidt, L., Ritter, H.: Computational models of visual tagging. In: Rickheit, G., Wachsmuth, I. (eds.) Situated Communication, pp. 209–242. De Gruyter, Berlin (2006)
8. Rayner, K.: Understanding eye movements in reading. Scientific Studies of Reading 1, 317–339 (1997)
9. Rettie, R., Brewer, C.: The verbal and visual components of package design. Journal of Product and Brand Management 9, 56–70 (2000)
10. Stampe, D.: Heuristic filtering and reliable calibration methods for video-based pupil-tracking systems. Behavioral Research Methods, Instruments and Computers 25, 137–142 (1993)
11. Underwood, G.: Eye fixations on pictures of natural scenes: getting the gist and identifying the components. In: Underwood, G. (ed.) Cognitive Processes in Eye Guidance, pp. 163–187. Oxford University Press, Oxford (2005)
12. Vertegaal, R.: Designing attentive interfaces. In: Proceedings of the symposium on eye tracking research and applications (ETRA 2002), pp. 22–30 (2002)
13. Zhai, S.: What's in the eyes for attentive input. Communication ACM 46, 34–39 (2003)

Integrating Inhomogeneous Processing and Proto-object Formation in a Computational Model of Visual Attention

Marco Wischnewski, Jochen J. Steil, Lothar Kehrer, and Werner X. Schneider

Abstract. We implement a novel computational framework for attention that includes recent experimentally derived assumptions on attention which are not covered by standard computational models. To this end, we combine inhomogeneous visual processing, proto-object formation, and parts of TVA (Theory of Visual Attention [2]), a well established computational theory in experimental psychology, which explains a large range of human and monkey data on attention. The first steps of processing employ inhomogeneous processing for the basic visual feature maps. Next, we compute so-called proto-objects by means of blob detection based on these inhomogeneous maps. Our model therefore displays the well known "global-effect" of eye movement control, that is, saccade target landing objects tend to fuse with increasing eccentricity from the center of view. The proto-objects also allow for a straightforward application of TVA and its mechanism to model task-driven selectivity. The final stage of our model consists of an attentional priority map which assigns priority to the proto-objects according to the computations of TVA. This step allows to restrict sophisticated filter computation to the proto-object regions and thereby renders our model computationally efficient by avoiding a complete standard pixel-wise priority computation of bottom-up saliency models.

Marco Wischnewski
Center of Excellence - Cognitive Interaction Technology (CITEC) and
Neuro-cognitive Psychology, Bielefeld University
e-mail: marco.wischnewski@cit-ec.uni-bielefeld.de

Jochen J. Steil
Research Institute for Cognition and Robotics (CoR-Lab), Bielefeld University
e-mail: jsteil@cor-lab.uni-bielefeld.de

Lothar Kehrer
Neuro-cognitive Psychology, Bielefeld University
e-mail: lothar.kehrer@uni-bielefeld.de

Werner X. Schneider
Neuro-cognitive Psychology and Center of Excellence - Cognitive Interaction Technology (CITEC), Bielefeld University
e-mail: wxs@uni-bielefeld.de

H. Ritter et al. (Eds.): Human Centered Robot Systems, COSMOS 6, pp. 93–102.
springerlink.com

1 Introduction

With the advent of increasingly cognitive robots using active vision, computational models for guidance of their visual attention become a standard component in their control architectures [1, 6, 12, 16]. Many of these models are driven by the ambition to realize human-like perception and action and rely on bottom-up computation of features. The most influential model has been proposed by Itti & Koch [10]. It computes a pixel-wise saliency map that shifts attention to the location with the highest saliency value. However, current computational models are not able to explain classical human data on covert visual attention (e.g., from "visual search" or "partial report paradigms") with the same degree of specification and predictive power as psychological models [2, 4, 15, 18, 21]. In this paper, we argue that progress is possible by relying on one of the most sophisticated theoretical frameworks of experimental psychology and cognitive neuroscience, the Theory of Visual Attention (TVA, developed by Bundesen [2]). TVA explains essential empirical findings of human visual attention [3] and has a neurophysiological interpretation fitting findings from all major single cell studies on attention [4]. TVA can serve as a pivotal element in computational modeling of visual attention because it both allows for simple weighting of low level feature channels and proposes mechanisms for object-based task-driven control of processing resources.

TVA differs from most existing bottom-up saliency models in proposing that saliency (priority) is not computed pixel-wise, but rather entity-wise based on perceptual units. These units are competing elements of an attention priority map (APM). While TVA itself does not specify how these units are formed, we will assume that the perceptual units can be described as so-called proto-objects. Proto-objects are formed within the APM and they refer to homogeneous regions in low-level feature maps, which can be detected without sophisticated object recognition. There are some other recent models which rely on proto-object formation [13, 17], but not in connection with TVA. One extension of the classical Itti & Koch model [19] forms proto-objects around the maxima of the saliency map. In contrast to our model, it assumes that saliency has been already determined before forming the proto-object postattentively.

Our model gains further biological and psychological plausibility by implementing inhomogeneous low-level feature processing which is lacking in most recent models (e.g. [19]). It is based on detailed findings about processing of retinal information in early stages of the visual cortex [20]. In contrast, most standard bottom-up attention-models operate pixel-wise in the visual field and it makes no difference whether an object (or a feature) appears foveally or in the periphery. Therefore, they cannot explain classical effects that demonstrate the inhomogeneous nature of visual processing such as the well-known "global effect" [7] of eye movement control. Saccadic eye movements to two nearby objects tend to land within the center-of-gravity of these objects given eye movements have to be made under time pressure. Given spatial proximity of the two objects, our model computes in this case one common proto-object. Importantly, this averaging effect increases spatially with increasing retinal eccentricity. Due to the inhomogeneity of the feature maps, our model shows this effect because proto-object computation in the periphery allows fewer candidate regions (proto-objects) to survive as individuated single objects.

Our model implements the whole path from visual input up to the APM (see Fig. 1). Incipient with the input image, we compute the inhomogeneous feature maps for color and luminance of one selected filter level for determining the proto-objects. This selection is motivated by the finding in human experiments which suggest that under time pressure only filters up to a certain resolution level enter the computation [11]. At this stage, costly computation of all Gabor filters is still delayed until information about the proto-objects regions is available. Further, only for the proto-object regions all filters are computed and summed according to the TVA equations, which allow for a task specific weighting of feature channels. Based on these computations, the attentional priority map (APM) according to TVA is

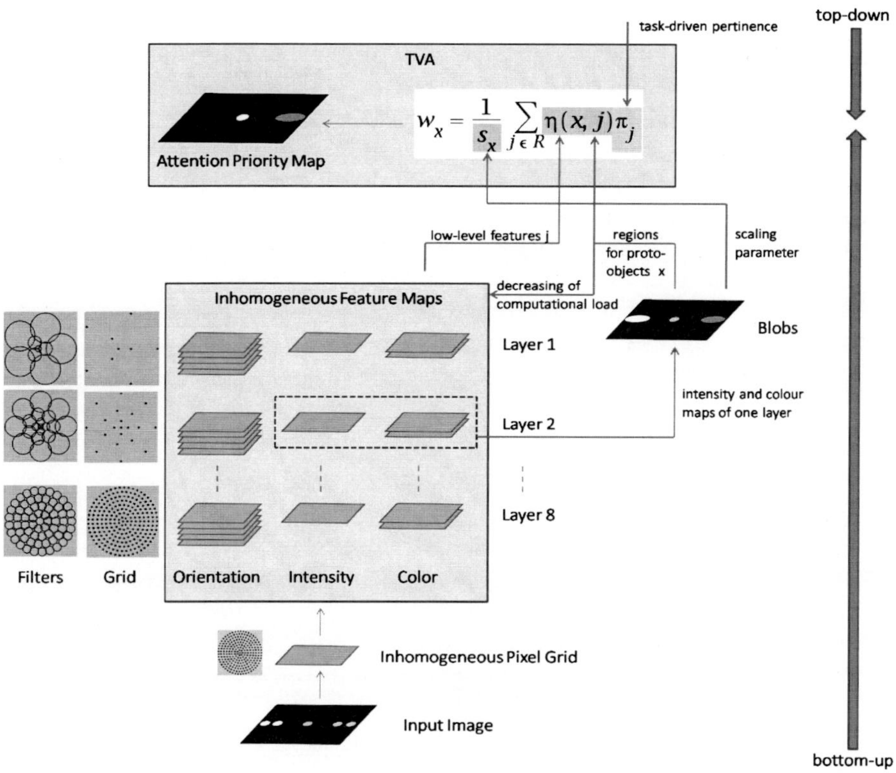

Fig. 1 The figure illustrates the structure of the model. A TVA-driven attention priority map (APM) results from top-down (pertinence) and bottom-up (proto-objects and feature maps) processes. In this example, the peripherally located proto-objects show the global effect. Furthermore, the proto-object on the left side disappears in the APM due to being task-irrelevant. Finally, the attentional weight of the proto-object on the right side is downscaled according to its high angle of eccentricity.

formed. We assume that the APM serves as linkage between the ventral ("what") and the dorsal ("where") pathway. Proto-object computation is performed by the dorsal pathway while attentional priorities for proto-objects are computed within the ventral pathway (e.g. [15]). The subsequent sections guide along the path illustrated in Fig. 1, incipient with the input image up to the APM. Examples illustrate the properties of our model in terms of the global-effect.

2 Inhomogeneous Retinal/V1 Processing

To comply with the inhomogeneous density of photoreceptors in the human retina [14], the homogeneous pixel grid input image (e.g. from a robot's camera) is transformed into an inhomogeneous pixel grid which serves as input for all subsequent filter operations (see Fig. 2, left). The grid positions ("receptors") are computed in the same way as the positions of the subsequent filters, but to cope with the Nyquist-Theorem, the density is doubled in relation to the filter layer with the highest density. The inhomogeneous feature maps are based on a biological driven mathematical description of V1 bar and edge Gabor filters developed by Watson [20]. The inhomogeneity of the filter structure is defined by the following relations: With increasing angle of eccentricity (a) the filters' size and (b) the distance between adjacent filters increases, whereas (c) the filters' spatial frequency decreases. The scaling s is linear with respect to a scaling parameter k (see. Eq. (1)), where e is the angle of eccentricity in degree. In the human visual system, k is estimated around 0.4 [20].

$$s = 1 + k * e \tag{1}$$

According to Watson, all subsequent parameters are computed as follows: At the center of the visual field, with $e = 0$, there is a central filter with $s = 1$. This filter

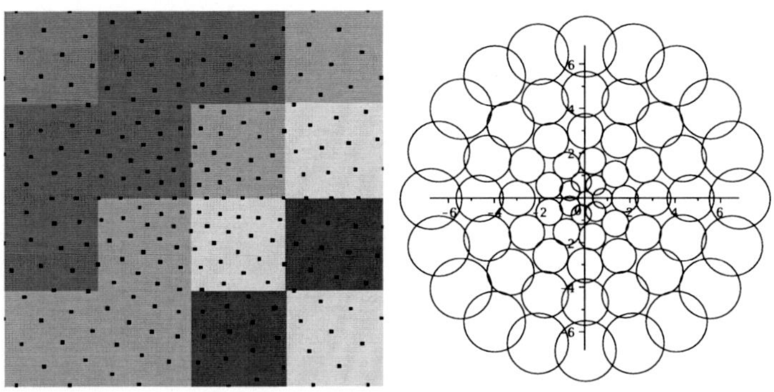

Fig. 2 Left: The figure shows an input image centric section of 4x4 pixel. The black dots match the inhomogeneous pixel grid positions. Right: The first five rings surrounding the central filter with $f_{center} = 1$ and $k = 0.4$. The axes of abscissae and ordinate reflect the angle of eccentricity. For illustration, the filter's size was reduced.

has a given central spatial frequency f_{center}. The size of the filter is assigned by its width at half height: $w_{center} = 1.324/f_{center}$. With increasing eccentricity, concentrical rings of filters are generated recursively. The parameters of an inner ring (or the central filter) determine the distance to the next outer ring as well as the distance (and thereby the number) of the filters within this outer ring: $d = 1/(f * 1.33)$. The frequency and width for each filter are adjusted as $f = f_{center}/s$ and $w = w_{center} * s$. Consequently, there can be added as many rings as necessary to cover a desired area of the visual field (see Fig. 2, right).

Each Gabor filter consists of a cosine function overlaid by a Gaussian function where θ represents the angle and ϕ the phase (2).

$$f(x,y) = \exp^{\frac{4\ln(2)(x^2+y^2)}{w^2}} cos(2\pi f(x\cos\theta + y\sin\theta) + \phi) \tag{2}$$

For each filter, the orientation is varied fivefold (0°, 36°, 72°, 108° and 144°) and the phase twice (0° and 90°). For each orientation, the filters' outcome of both phases (bar and edge filter) is combined to obtain the locally shift invariant output [9].

This results in five orientation feature maps (each represents one orientation) for one filter structure given a central frequency. To cover the whole human frequency space, it is necessary to layer these structures. Thus, the model consists of 8 layers in which the first layer starts with a center frequency of 0.25 cyc/deg. From layer n to layer $n+1$, this center frequency is doubled, so that at the end layer eight has a center frequency of 32 cyc/deg. We obtain 40 feature maps (8 layer each with 5 orientation feature maps). Note that, however, the full filter bank of Gabor filters needs to be computed only for grid position comprising proto-objects, which are determined based on the color and intensity filters for a certain selected resolution alone.

For the color and intensity feature maps, the described filter structure is adopted, but filters are restricted to the Gaussian part in (2). The color feature maps rely on the physiological RG/BY space [19]. Again, 8 layers are computed, so there are 8 intensity and 16 color feature maps (8 RG and 8 BY). In sum, we obtain 64 feature maps. Again note that they have to compute in full scale only for the proto-object regions.

3 Proto-object Formation

For blob detection we use an algorithm developed by Forssén [8]. It makes use of channel representations within the three-dimensional color/intensity space and thereby allows for spatial overlapping of resultant blobs, which are spatial homogeneous regions approximated by ellipses. The intensity and color feature maps of one layer serve as input. The choice of the layer simulates to what extend the system is under time pressure, because high-resolution layers need more time for processing. Thus, a high degree of time pressure yields a low-resolution layer as input and thereby merging of objects into one proto-object is observable (global-effect). In order to utilize the blob algorithm, the inhomogeneous pixel grid of the feature maps is transformed back into voronoi cells on a homogeneous pixel grid (see Fig. 3, c). The size of the homogeneous pixel grid is, layer-independent and constant, as large as being necessary to avoid an overlapping of back transformed pixel even if the layer with highest center frequency was chosen.

Due to the peripherally increasing size of the voronoi cells, we included a filtering mechanism. That is, every blob has to have at least a size of $n * v$ in both of its axes where v is the size of the voronoi cell containing the blob's centroid and n is a scaling factor with $n > 1$. Thus, depending on factor n, a minimum number of n^2 cohering voronoi cells are necessary to build a blob, which works uncoupled with respect to the angle of eccentricity.

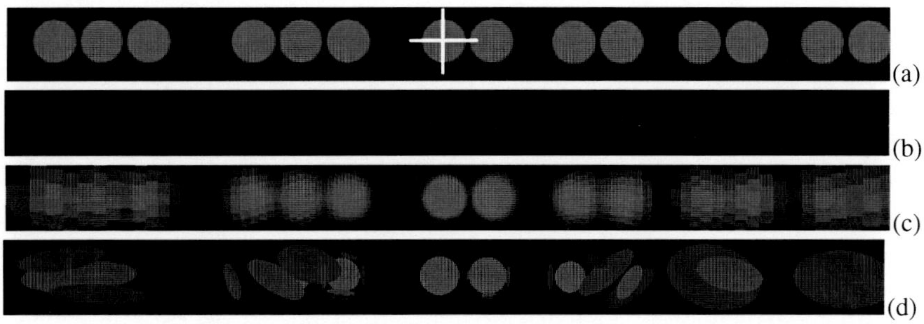

Fig. 3 The figure shows (a) the input image in which the white cross marks the foveal center, (b) the inhomogeneous pixel grid, (c) the voronoi cells, and (d) the blobs. To demonstrate the global effect, circles combined to groups of two and three are used. The blobs in (d) show the change from the fovea to the periphery: Whereas foveally positioned blobs represent rather accurate the circles of the input image, peripherally positioned blobs tend to represent more circles and to be more inaccurately. The intensity of each blob reflects the average intensity of the channel's region.

4 TVA

In TVA, each proto-object x within the visual field has a weight which is the outcome of the *weight equation* (3). A spatial structured representation of all these weights is called the *attentional priority map* (APM). Each w_x-value is computed as the sum over all features which are element of R. The $\eta(x, j)$-value indicates the degree of proto-object x having feature j weighted by top-down task-dependent controlled pertinence π_j (e.g. search for a red object). Thus the $\eta(x, j)$-value restricts feature computation to proto-object regions, while π_j implement a standard feature channel weighting as also present in other saliency models.

$$w_x = \sum_{j \in R} \eta(x, j)\pi_j \tag{3}$$

At this point, all needed bottom-up data are available to compute the η-values: The region of each proto-object as found by the blob detection algorithm, and the features computed within this region as determined in the inhomogeneous feature maps.

If two proto-objects p_1 and p_2 are completely equal in the size of their regions and the values of the feature maps within these regions and vary only in the angle of eccentricity, then the more foveally located proto-object gets a higher attentional weight. A computational expedient way to integrate this relation is the implementation of an *inhomogeneity parameter* s_x which represents the scaling of proto-object x depending on its angle of eccentricity (4, right). The mathematical embedding of s_x leads to a modified weight equation (4, left). Thus, if it is the case that $s_1 = 2s_2$, then p_1's region has to be double in size to get the same attentional weight as p_2.

$$w_x = \frac{1}{s_x} \sum_{j \in R} \eta(x,j)\pi_j \quad with \quad s_x = 1 + k * e_x \tag{4}$$

5 Results

How does the global effect emerge from our computational architecture? Fig. 5 shows the result of the blob-algorithm to determine the proto-objects of layer 2 to 8 depending on the hyperparameter k. Layer one was skipped because it produces no blobs. Layers with lower frequency, whose choice was motivated by time pressure for saccadic eye movements, produce the global effect. This means, the visual system cannot distinguish adjacent objects in consequence of the low spatial filter resolution. Therefore, these objects fuse together to a "surrounding" proto-object and a saccadic eye movement lands in the center-of-gravity of these objects within the visual field which roughly equals the center of this proto-object.

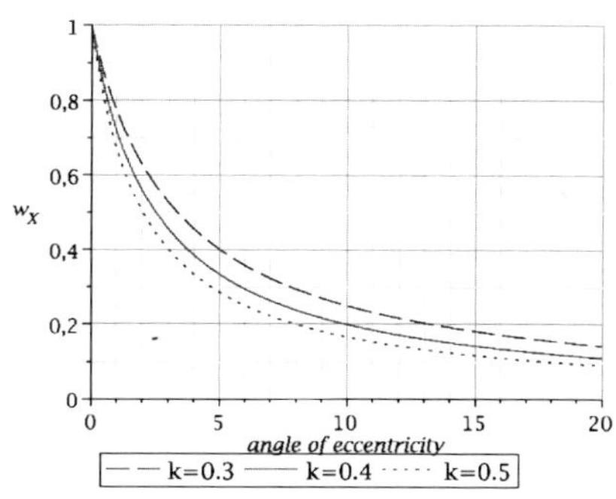

Fig. 4 The figure illustrates the influence of the inhomogeneous parameter s_x on the attentional weight w_x for different k-values. The axis of abscissae reflects the angle of eccentricity and the axis of ordinates the attentional weight w_x.

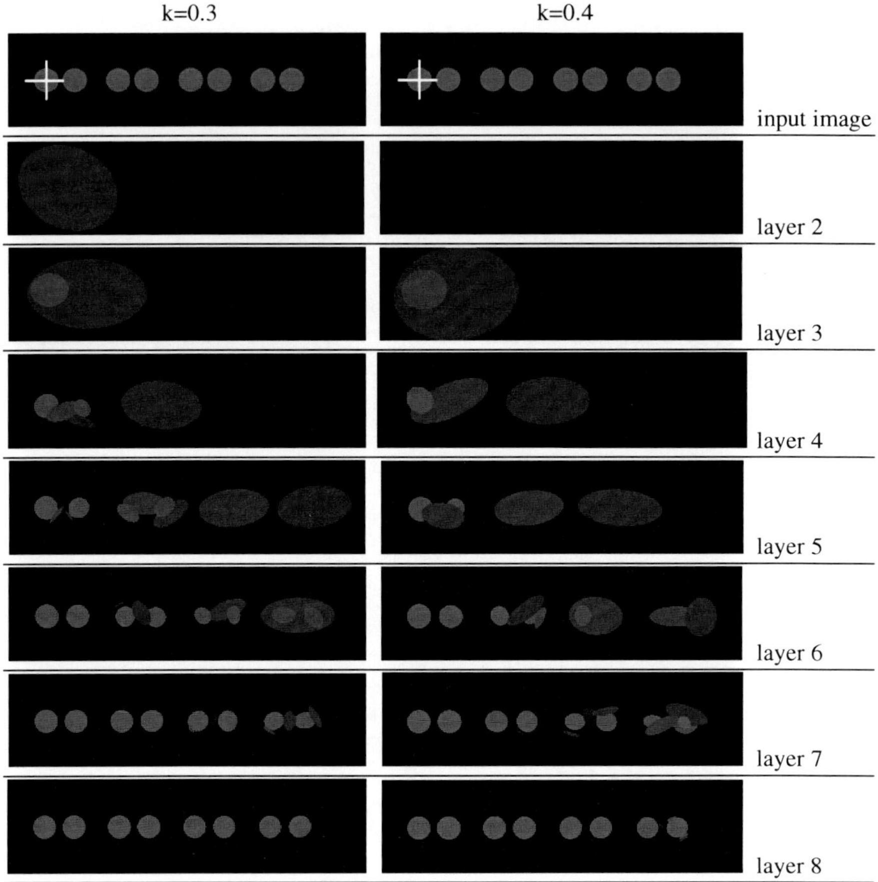

Fig. 5 The figure shows the resulting blobs depending on feature layer and parameter k. On the top, the input image is shown. The white cross marks the foveal center and the intensity of each blob reflects the average intensity of the channel's region.

Layers with lower frequency also produce, due to the lower spatial resolution, larger blobs. The lower the frequency, the more objects within the periphery of the visual field are not represented by a proto-object. They simply disappear. The k parameter is a hyperparameter for scaling these effects within a layer, because decreasing the k-value leads, according to the scaling function (1), to a slower decrease of the filter density from the foveal center to the periphery. We choose k motivated by the finding in the human visual system [20].

Applied to images consisting of natural objects our model yields psychological feasible results (see Fig. 6). Peripherally located objects tend to be represented by only one proto-object or even disappear, whereas more foveally located objects tend to produce proto-objects which represent parts of them.

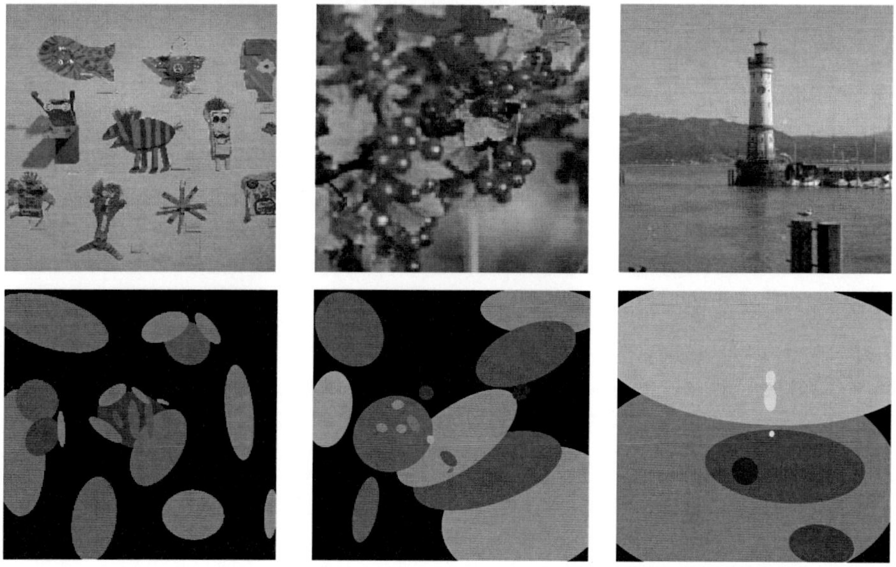

Fig. 6 Proto-objects formed on the basis of inhomogeneous processing in natural images with $f_{center} = 4$ and $k = 0.4$

Finally, Fig. 4 shows the influence of the inhomogeneous parameter s_x on the attentional weight w_x. If we assume a proto-object with $\sum \eta(x, j) * \pi_j = 1$, the figure illustrates the outcome of the modified TVA weight equation (4) which then only depends on s_x. This is shown for different k-values. The higher the k-value, the stronger the angle of eccentricity affects the attentional weights.

6 Outlook

We attempted to show in this paper that the combination of inhomogeneous processing, proto-object formation and TVA leads to new and interesting forms of controlling overt and covert visual attention [5]. Moreover, proto-object computation allows to include sophisticated task-driven control of visual attention according to TVA. Furthermore, this approach provides the possibility to reduce computational load. Only those features from the feature maps (orientation, intensity and color) which are located in proto-object regions have to be computed for the η-values of TVA. These bottom-up η-values are multiplicatively combined with top-down pertinence (task) values and result in proto-object based attentional weights. Future computational research is needed in order to evaluate the potential of this TVA-based task-driven form of attentional control for robotics, psychology and cognitive neuroscience.

References

1. Breazeal, C., Scassellati, B.: A context-dependent attention system for a social robot. In: Proc. 16th IJCAI, pp. 1146–1153 (1999)
2. Bundesen, C.: A theory of visual attention. Psych. Rev. 97, 523–547 (1990)
3. Bundesen, C., Habekost, T.: Principles of visual attention. Oxford University Press, Oxford (2008)
4. Bundesen, C., Habekost, T., Killingsbaek, S.: A neural theory of visual attention: Bridging cognition and neurophysiology. Psych. Rev. 112, 291–328 (2005)
5. Deubel, H., Schneider, W.X.: Saccade target selection and object recognition: Evidence for a common attentional mechanism. Vis. Res. 36, 1827–1837 (1996)
6. Driscoll, J.A., Peters II, R.A., Cave, K.R.: A visual attention network for a humanoid robot. In: Proc. IEEE/RSJ IROS 1998, pp. 12–16 (1998)
7. Findlay, J.M.: Global processing for saccadic eye movements. Vis. Res. 22, 1033–1045 (1982)
8. Forssén, P.E.: Low and medium level vision using channel representations. Dissertation No. 858 (2004), ISBN 91-7373-876-X
9. Fogel, I., Sagi, D.: Gabor filters as texture dicriminator. Biol. Cybern. 61, 103–113 (1989)
10. Itti, L., Koch, C., Niebur, E.: A model of saliency-based visual attention for rapid scene analysis. In: IEEE ICCV, pp. 195–202 (1998)
11. Kehrer, L.: Central performance drop on perceptual segregation tasks. Spatial Vision 4, 45–62 (1989)
12. Nagai, Y., Hosoda, K., Morita, A., Asada, M.: A constructive model for the development of joint attention. Connection Science 15, 211–229 (2003)
13. Orabona, F., Metta, G., Sandini, G.: A Proto-object based visual attention model. In: Paletta, L., Rome, E. (eds.) WAPCV 2007. LNCS (LNAI), vol. 4840, pp. 198–215. Springer, Heidelberg (2007)
14. Palmer, S.E.: Vision Science, pp. 29–31. The MIT Press, Cambridge (1999)
15. Schneider, W.X.: VAM: A neuro-cognitive model for visual attention control of segmentation, object recognition, and space-based motor action. Vis. Cog. 2, 331–375 (1995)
16. Steil, J.J., Heidemann, G., Jockusch, J., Rae, R., Jungclaus, N., Ritter, H.: Guiding attention for grasping tasks by gestural instruction: The GRAVIS-robot architecture. In: Proc. IEEE/RSJ IROS 2001, pp. 1570–1577 (2001)
17. Sun, Y., Fisher, R., Wang, F., Gomes, H.M.: A computer vision model for visual-object-based attention and eye movements. Computer Vision and Image Understanding 112, 126–142 (2008)
18. Treisman, A.M.: Features and objects: The fourteenth Bartlett memorial lecture. Quarterly Journal of Experimental Psychology 40A, 201–237 (1988)
19. Walther, D., Koch, C.: Modeling attention to salient proto-objects. Neural Networks 19, 1395–1407 (2006)
20. Watson, A.B.: Detection and recognition of simple spatial forms. In: Braddick, O.J., Sleigh, A.C. (eds.) Physiological and biological processing of images, pp. 100–114. Springer, Heidelberg (1983)
21. Wolfe, J.M.: Guided search 2.0: a revised model of visual search. Psychonomic Bulletin and Review 1, 202–238 (1994)

Dimensionality Reduction in HRTF by Using Multiway Array Analysis

Martin Rothbucher, Hao Shen, and Klaus Diepold

Abstract. In a human centered robotic system, it is important to provide the robotic platform with multimodal human-like sensing, e.g. haptic, vision and audition, in order to improve interactions between the human and the robot. Recently, Head Related Transfer Functions (HRTFs) based techniques have become a promising methodology for robotic binaural hearing, which is a most prominent concept in human robot communication. In complex and dynamical applications, due to its high dimensionality, it is inefficient to utilize the originial HRTFs. To cope with this difficulty, Principle Component Analysis (PCA) has been successfully used to reduce the dimensionality of HRTF datasets. However, it requires in general a vectorization process of the original dataset, which is a three-way array, and consequently might cause loss of structure information of the dataset. In this paper we apply two multiway array analysis methods, namely the Generalized Low Rank Approximations of Matrices (GLRAM) and the Tensor Singular Value Decomposition (Tensor-SVD), to dimensionality reductions in HRTF based applications. Our experimental results indicate that an optimized GLRAM outperforms significantly the PCA and performs nearly as well as Tensor-SVD with less computational complexity.

1 Introduction

Head Related Transfer Functions (HRTFs) describe spectral changes of sound waves when they enter the ear carnal, due to the diffraction and reflection properties of the human body, i.e. the head, shoulders, torso and ears. In far field applications, they can be considered as complicated functions of frequency and two spatial variables (elevation and azimuth) [2]. Thus HRTFs can be considered as direction dependent filters. Since the geometric features of head, shoulders, torso and ears differ from person to person, HRTFs are unique for each individual.

Martin Rothbucher · Hao Shen · Klaus Diepold
Institute for Data Processing, Technische Universität München, 80290 München, Germany
e-mail: {martin.rothbucher,hao.shen,kldi}@tum.de

H. Ritter et al. (Eds.): Human Centered Robot Systems, COSMOS 6, pp. 103–110.
springerlink.com

The human cues of sound localization can be used in telepresence applications, where humans control remote robots (e.g. for dangerous tasks). Thus the sound, localized by the robot, is then synthesized at the human's site in order to reconstruct the auditory space around the robot. Robotic platforms would benefit from a human based sound localization approach because of its noise-tolerance and the ability to localize sounds in a three-dimensional environment with only two microphones, or reproducing 3D sound with simple headphones. Consequently, it is of great use to improve performance of sound localization and synthesis for telepresence systems.

Recently, researchers have invested great efforts in customization of HRTFs [4], which leads to better quality of sound synthesis with respect to unique individuals [12]. However, for the purpose of HRTF customization, a huge number of Head Related Impulse Response (HRIR) datasets of various test subjects are usually required. Therefore in order to make a customization application more efficient, even on a computational restricted system, e.g. telepresence systems or mobile phones, dimensionality reduction methods can be used.

Since the pioneering paper [6], Principle Component Analysis (PCA) has become a prominent tool for HRTF reduction [8] and customization [4]. In general, applying PCA to HRTF datasets requires a vectorization process of the original 3D dataset. As a consequence, some structure information of the HRTF dataset might be lost. To avoid such limits, the so-called Tensor-SVD method, which was originally introduced in the community of multiway array analysis [7], has been recently applied into HRTF analysis, in particular for customization [3]. Meanwhile, in the community of image processing, the so-called Generalized Low Rank Approximations of Matrices (GLRAM) [13], a generalized form of 2DPCA, has been developed in competition with the standard PCA. It has been shown, that the GLRAM method is essentially a simple form of Tensor-SVD [10]. In this paper, we study both GLRAM and Tensor-SVD methods for the purpose of dimensionality reduction of HRIR datasets and compare them with the standard PCA.

The paper is organized as follows. Section 2 provides a brief introduction to three dimensionality reduction methods, namely, PCA, GLRAM and Tensor-SVD. In section 3, performance of the three methods is investigated by several numerical experiments. Finally, a conclusion is given in section 4.

2 HRIR Reduction Techniques

In this section, we briefly describe three techniques of dimensionality reduction for HRIRs, namely, PCA, GLRAM and Tensor-SVD. In general, each HRIR dataset can be represented as a three-way array $\mathscr{H} \in \mathbb{R}^{N_a \times N_e \times N_t}$, where the dimensions N_a and N_e are the spatial resolutions of azimuth and elevation, respectively, and N_t the time sample size. By a Matlab-like notation, in this work we denote $\mathscr{H}(i,j,k) \in \mathbb{R}$ the (i,j,k)-th entry of $\mathscr{H}$, $\mathscr{H}(l,m,:) \in \mathbb{R}^{N_t}$ the vector with a fixed pair of (l,m) of $\mathscr{H}$ and $\mathscr{H}(l,:,:) \in \mathbb{R}^{N_e \times N_t}$ the l-th slide (matrix) of $\mathscr{H}$ along the azimuth direction.

2.1 Principal Component Analysis

The dimensionality reduction of HRIRs by using PCA is described as follows. First of all, we construct the matrix

$$H := \left[\mathrm{vec}(\mathscr{H}(:,:,1))^\top, \ldots, \mathrm{vec}(\mathscr{H}(:,:,N_t))^\top\right] \in \mathbb{R}^{N_t \times (N_a \cdot N_e)}, \tag{1}$$

where the operator $\mathrm{vec}(\cdot)$ puts a matrix into a vector form. Let $H = [h_1, \ldots, h_{N_t}]$. The mean value of columns of H is then computed by

$$\mu = \tfrac{1}{N_t} \sum_{i=1}^{N_t} h_i. \tag{2}$$

After centering each row of H, i.e. computing $\widehat{H} = [\widehat{h}_1, \ldots, \widehat{h}_{N_t}] \in \mathbb{R}^{N_t \times (N_a \cdot N_e)}$ where $\widehat{h}_i = h_i - \mu$ for $i = 1, \ldots, N_t$, the covariance matrix of $\widehat{H}$ is computed as follows

$$C := \tfrac{1}{N_t} \widehat{H} \widehat{H}^\top. \tag{3}$$

Now we compute the eigenvalue decomposition of C and select q eigenvectors $\{x_1, \ldots, x_q\}$ corresponding to the q largest eigenvalues. Then by denoting $X = [x_1, \ldots, x_q] \in \mathbb{R}^{N_t \times q}$, the HRIR dataset can be reduced by the following

$$\widetilde{H} = X^\top \widehat{H} \in \mathbb{R}^{q \times (N_a \cdot N_e)}. \tag{4}$$

Note, that the storage space for the reduced HRIR dataset depends on the value of q. Finally to reconstruct the HRIR dataset one need to simply compute

$$H_r = X \widetilde{H} + \mu \in \mathbb{R}^{N_t \times (N_a \cdot N_e)}. \tag{5}$$

We refer to [5] for further discussions on PCA.

2.2 Tensor-SVD of Three-Way Array

Unlike the PCA algorithm vectorizing the HRIR dataset, Tensor-SVD keep the structure of the original 3D dataset intact. Given a HRIR dataset $\mathscr{H} \in \mathbb{R}^{N_a \times N_e \times N_t}$, Tensor-SVD computes its best multilinear *rank* $-(r_a, r_e, r_t)$ approximation $\widehat{\mathscr{H}} \in \mathbb{R}^{N_a \times N_e \times N_t}$, where $N_a > r_a$, $N_e > r_e$ and $N_t > r_t$, by solving the following minimization problem

$$\min_{\widehat{\mathscr{H}} \in \mathbb{R}^{N_a \times N_e \times N_t}} \left\| \mathscr{H} - \widehat{\mathscr{H}} \right\|_{\mathrm{F}}, \tag{6}$$

where $\|\cdot\|_{\mathrm{F}}$ denotes the Frobenius norm of tensors. The *rank* $-(r_a, r_e, r_t)$ tensor $\widehat{\mathscr{H}}$ can be decomposed as a *trilinear* multiplication of a *rank* $-(r_a, r_e, r_t)$ core tensor $\mathscr{C} \in \mathbb{R}^{r_a \times r_e \times r_t}$ with three full-rank matrices $X \in \mathbb{R}^{N_a \times r_a}$, $Y \in \mathbb{R}^{N_e \times r_e}$ and $Z \in \mathbb{R}^{N_t \times r_t}$, which is defined by

$$\widehat{\mathscr{H}} = (X,Y,Z) \cdot \mathscr{C} \tag{7}$$

where the (i,j,k)-th entry of $\widehat{\mathscr{H}}$ is computed by

$$\widehat{\mathscr{H}}(i,j,k) = \sum_{\alpha=1}^{r_a} \sum_{\beta=1}^{r_e} \sum_{\gamma=1}^{r_t} x_{i\alpha} y_{j\beta} z_{k\gamma} \mathscr{C}(\alpha,\beta,\gamma). \tag{8}$$

Thus without loss of generality, the minimization problem as defined in (6) is equivalent to the following

$$\begin{aligned} &\min_{X,Y,Z,\mathscr{C}} \|\mathscr{H} - (X,Y,Z) \cdot \mathscr{C}\|_{\mathrm{F}}, \\ &\text{s.t.} \quad X^\top X = I_{r_a}, Y^\top Y = I_{r_e} \text{ and } Z^\top Z = I_{r_t}. \end{aligned} \tag{9}$$

We refer to [9] for Tensor-SVD algorithms and further discussions.

2.3 Generalized Low Rank Approximations of Matrices

Similar to Tensor-SVD, GLRAM methods does not require destruction of a 3D tensor. Instead of compressing along all three directions as Tensor-SVD, GLRAM methods work with two pre-selected directions of a 3D data array.

Given a HRIR dataset $\mathscr{H} \in \mathbb{R}^{N_a \times N_e \times N_t}$, we assume to compress $\mathscr{H}$ in the first two directions. Then the task of GLRAM is to approximate slides (matrices) $\mathscr{H}(:,:,i)$, for $i = 1,\ldots,N_t$, of $\mathscr{H}$ along the third direction by a set of low rank matrices $\{XM_iY^\top\} \subset \mathbb{R}^{N_a \times N_e}$, for $i = 1,\ldots,N_t$, where the matrices $X \in \mathbb{R}^{N_a \times r_a}$ and $Y \in \mathbb{R}^{N_e \times r_e}$ are of full rank, and the set of matrices $\{M_i\} \subset \mathbb{R}^{r_a \times r_e}$ with $N_a > r_a$ and $N_e > r_e$. This can be formulated as the following optimization problem

$$\begin{aligned} &\min_{X,Y,\{M_i\}_{i=1}^{N_t}} \sum_{i=1}^{N_t} \left\| (\mathscr{H}(:,:,i) - XM_iY^\top) \right\|_{\mathrm{F}}, \\ &\text{s.t.} \quad X^\top X = I_{r_a} \text{ and } Y^\top Y = I_{r_e}. \end{aligned} \tag{10}$$

Here, by abuse of notations, $\|\cdot\|_{\mathrm{F}}$ denotes the Frobenius norm of matrices. Let us construct a 3D array $\mathscr{M} \in \mathbb{R}^{r_a \times r_e \times N_t}$ by assigning $\mathscr{M}(:,:,i) = M_i$ for $i = 1,\ldots,N_t$. The minimization problem as defined in (10) can be reformulated in a Tensor-SVD style, i.e.

$$\begin{aligned} &\min_{X,Y,\mathscr{M}} \|\mathscr{H} - (X,Y,I_{N_t}) \cdot \mathscr{M}\|_{\mathrm{F}}, \\ &\text{s.t.} \quad X^\top X = I_{r_a} \text{ and } Y^\top Y = I_{r_e}. \end{aligned} \tag{11}$$

We refer to [13] for more details on GLRAM algorithms.

Remark 0.1. GLRAM methods work on two pre-selected directions out of three. There are then in total three different combinations of directions to implement GLRAM on an HRIR dataset. Performance of GLRAM in different directions might vary significantly. This issue will be investigated and discussed in section 3.2.1.

3 Numerical Simulations

In this section, we apply PCA, GLRAM and Tensor-SVD to a HRIR based sound localization problem, in order to investigate performance of these three methods for data reduction.

3.1 Experimental Settings

We use the CIPIC database [1] for our data reduction experiments. The database contains 45 HRIR tensors of individual subjects for both left and right ears, with a spatial resolution of 1250 points ($N_e = 50$ in elevation, $N_a = 25$ in azimuth) and the time sample size $N_t = 200$.

In our experiment, we use a convolution based algorithm [11] for sound localization. Given two signals S_L and S_R, representing the received left and right ear signals of a sound source at a particular location, the correct localization is expected to be achieved by maximizing the cross correlation between the filtered signals of S_L with the right ear HRIRs and the filtered signal of S_R with the left ear HRIRs.

3.2 Experimental Results

In each experiment, we reduce the left and right ear KEMAR HRIR dataset (subject 21 in the CIPIC database) with one of the introduced reduction methods. By randomly selecting 35 locations in the 3D space, a test signal, which is white noise in our experiments, is virtually synthesized using the corresponding original HRIRs. The convolution based sound localization algorithm, which is fed with the restored databases, is then used to localize the signals. Finally, the localization success rate is computed.

3.2.1 GLRAM in HRIR Dimensionality Reduction

In this section, we investigate performance of the GLRAM method to HRIR dataset reduction in three different combinations of directions. Firstly, we reduce HRIR datasets in the first two directions, i.e. azimuth and elevation. For a fixed pair of values (N_{r_a}, N_{r_e}), each pre-described experiment gives a localization success rate of the test signals. Then, for a given range of (N_{r_a}, N_{r_e}), the contour plot of localization

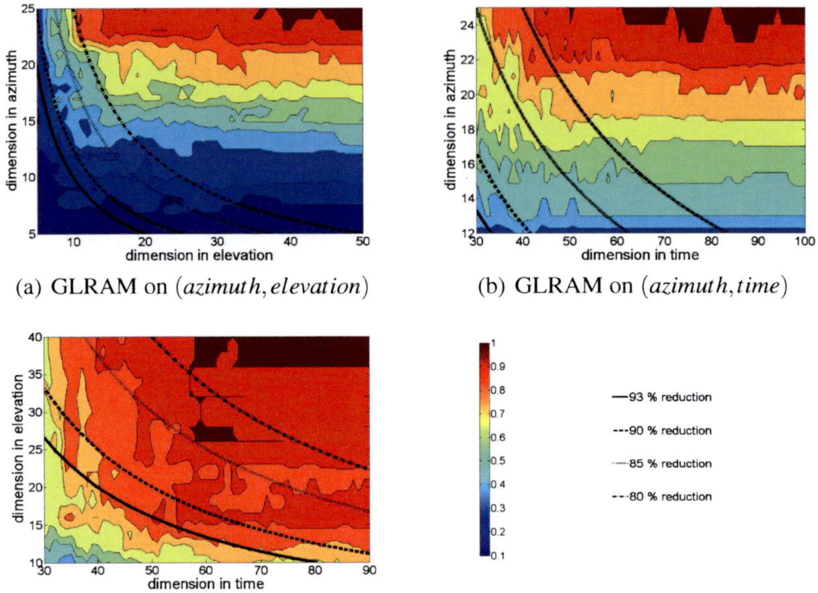

(a) GLRAM on $(azimuth, elevation)$

(b) GLRAM on $(azimuth, time)$

(c) GLRAM on $(elevation, time)$

Fig. 1 Contour plots of localization success rate of using GLRAM in different settings

success rate with respect to various pairs of (N_{r_a}, N_{r_e}) is drawn in Fig. 1(a). The curves in Fig.1(a) correspond to a set of fixed reduction rates. Similar results with respect to the pairs of $(azimuth, time)$ and $(elevation, time)$ are plotted in Fig. 1(b) and Fig. 1(c), respectively.

According to Fig. 1(a), to reach a success rate of 90%, the maximal reduction rate of 80% is achieved at $r_a = 23$ and $r_e = 13$. We then summarize similar results from Fig. 1(b) and Fig. 1(c) for different success rates in Table 1. It is obvious that

Table 1 Reduction rate achieved by using GLRAM at different localization success rates

Localization Success Rate	90%	80%	70%
GLRAM (N_{r_a}, N_{r_e})	80%	81%	82%
GLRAM (N_{r_a}, N_{r_t})	80%	83%	85%
GLRAM (N_{r_e}, N_{r_t})	93%	93%	94%

applying GLRAM on the pair of directions $(elevation, time)$ outperforms the other two combinations. A reduction on the azimuth direction, which has the smallest dimension, leads to great loss of localization accuracy. It might indicate that differences (Interaural Time Delays) presented in the HRIRs between neighboring azimuth angles have stronger influence on localization cues than differences between neighboring elevation angles.

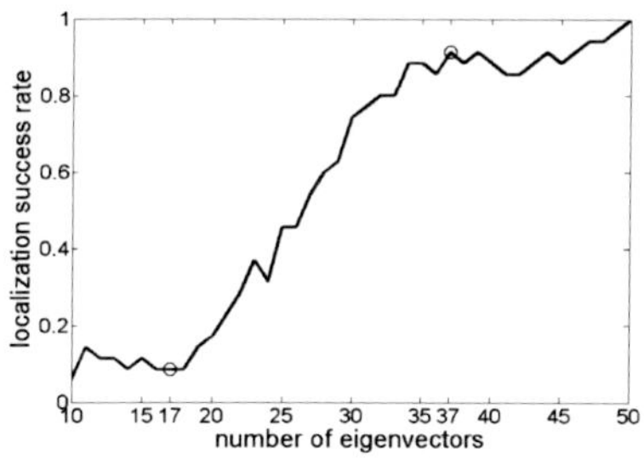

Fig. 2 success rate by PCA

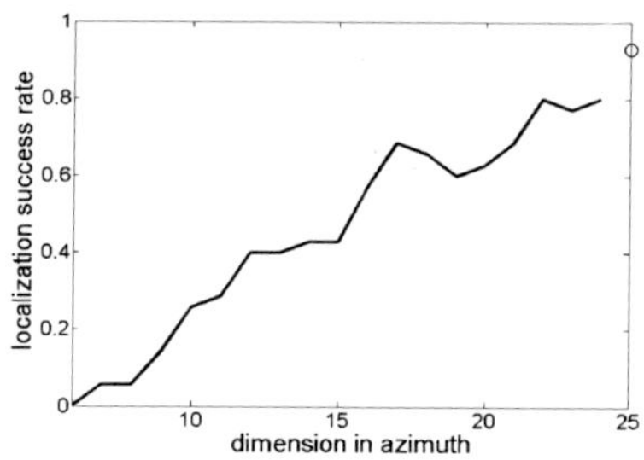

Fig. 3 success rate by Tensor-SVD

3.2.2 PCA and Tensor-SVD Reduced HRIRs

In the previous section, it is shown that the application of GLRAM in the directions of elevation and time performs the best, here we compare this optimal GLRAM with the standard PCA and Tensor-SVD.

First of all, localization success rate of the test signals by using PCA reduction with respect to the number of eigenvectors is shown in Fig. 2. It is known from Table 1 that, to reach the success rate of 90% with the optimal GLRAM, it achieves the maximal reduction rate of 93%, which is equivalent for PCA with 17 eigenvectors. It is shown in Fig. 2, that with 17 eigenvectors, a localization success rate of 9% is only reached. On the other hand, to reach the success rate of 90%, PCA requires 37 eigenvectors, which gives a reduction rate of 82%. It is thus clear that the optimal GLRAM outperforms the PCA remarkably.

As we know, GLRAM is a simple form of Tensor-SVD with leaving one direction out, in our last experiment, we investigate the effect of the third direction in performance of data reduction. We fix the dimensions in elevation and time to $r_e = 15$ and $r_t = 55$, which are the dimensions for the optimal GLRAM in achieving 90% success rate, see Fig. 1(c). Fig. 3 shows the localization success rate of using Tensor-SVD with a changing dimension in azimuth. The decrease of the dimension in azimuth leads to a consistently huge loss of localization accuracy. In other words, with fixed reductions in directions $(elevation, time)$, GLRAM outperforms Tensor-SVD. Similar to the observations in previous subsection, localization accuracy seems to be more sensitive to the reduction in the azimuth direction than the other two directions.

4 Conclusion and Future Work

In this paper, we address the problem of dimensionality reduction of HRIR dataset using PCA, Tensor-SVD and GLRAM. Our experiments demonstrate that an optimized GLRAM can beat the standard PCA reduction with a significant benefit. Meanwhile, GLRAM is also competitive to Tensor-SVD due to nearly equivalent performance and less computational complexity. Applying GLRAM on HRIR data, two possible applications for future work are recommended. First, in order to accelerate localization process on mobile robotic platforms, GLRAM methods can be used to shorten the HRTF filters. Secondly, GLRAM methods could also be beneficial in HRTF customization.

Acknowledgements. This work was fully supported by the German Research Foundation (DFG) within the collaborative research center SFB453 "High-Fidelity Telepresence and Teleaction".

References

1. Algazi, V.R., Duda, R.O., Thompson, D.M., Avendano, C.: The CIPIC HRTF database. In: IEEE ASSP Workshop on Applications of Signal Processing to Audio and Acoustics, New Paltz, NY, pp. 21–24 (2001)
2. Blauert, J.: An introduction to binaural technology. In: Gilkey, G., Anderson, T. (eds.) Binaural and Spatial Hearing, pp. 593–609. Lawrence Erlbaum, Hilldale (1997)
3. Grindlay, G., Vasilescu, M.A.O.: A multilinear (tensor) framework for HRTF analysis and synthesis. In: Proceedings of IEEE International Conference on Acoustics, Speech and Signal Processing, Honolulu, Hawaii, USA, vol. 1, pp. 161–164 (2007)
4. Hu, L., Chen, H., Wu, Z.: The estimation of personalized HRTFs in individual VAS. In: Proceedings of the 2008 Fourth International Conference on Natural Computation, Washington, DC, USA, pp. 203–207 (2008)
5. Jolliffe, I.T.: Principal Component Analysis, 2nd edn. Springer, Heidelberg (2002)
6. Kistler, D.J., Wightman, F.L.: A model of head-related transfer functions based on principal components analysis and minimum-phase reconstruction. Journal of the Acoustical Society of America 91(3), 1637–1647 (1992)
7. Lathauwer, L., Moor, B., Vandewalle, J.: A multilinear singular value decomposition. SIAM Journal on Matrix Analysis and Applications 21(4), 1253–1278 (2000)
8. Middlebrooks, J., Green, D.: Observations on a principal components analysis of head-related transfer functions. Journal of the Acoustical Society of America 92(1), 597–599 (1992)
9. Savas, B., Lim, L.: Best multilinear rank approximation of tensors with quasi-Newton methods on Grassmannians. Technical Report LITH-MAT-R-2008-01-SE, Department of Mathematics, Linköpings University (2008)
10. Sheehan, B.N., Saad, Y.: Higher order orthogonal iteration of tensors (HOOI) and its relation to PCA and GLRAM. In: Proceedings of the 2007 SIAM International Conference on Data Mining, Minnenpolis, Minnesota, USA, pp. 355–366 (2007)
11. Usman, M., Keyrouz, F., Diepold, K.: Real time humanoid sound source localization and tracking in a highly reverberant environment. In: Proceedings of 9th International Conference on Signal Processing, Beijing, China, pp. 2661–2664 (2008)
12. Wenzel, E., Arruda, M., Kistler, D., Wightman, F.: Localization using nonindividualized head-related transfer functions. Journal of the Acoustical Society of America 94, 111–123 (1993)
13. Ye, J.: Generalized low rank approximations of matrices. Machine Learning 61(1-3), 167–191 (2005)

Multimodal Laughter Detection in Natural Discourses

Stefan Scherer, Friedhelm Schwenker, Nick Campbell, and Günther Palm

Abstract. This work focuses on the detection of laughter in natural multiparty discourses. For the given task features of two different modalities are used from unobtrusive sources, namely a room microphone and a 360 degree camera. A relatively novel approach using Echo State Networks (ESN) is utilized to achieve the task at hand. Among others, a possible application is the online detection of laughter in human robot interaction in order to enable the robot to react appropriately in a timely fashion towards human communication, since laughter is an important communication utility.

1 Introduction

Paralinguistic dialog elements, such as laughter, moans and back channeling, are important factors of human to human interaction besides direct communication using speech. They are essential to convey information such as agreement or disagreement in an efficient and natural way. Furthermore, laughter is an indication for the positive perception of a discourse element by the laughing dialog partner, or an indication for uncertainty considering nervous or social laughters [1]. Overall laughter is a very communicative element of discourses that is necessary for "healthy" communication and it can be used to measure engagement in interaction [9, 16, 8, 10]. Lively discourses are not only important for face to face communication, but as we believe essential for the acceptance of artificial agents, such as robots or expert systems providing information using speech synthesis and other mainly human

Stefan Scherer · Friedhelm Schwenker · Günther Palm
Institute of Neural Information Processing, Ulm University
e-mail: {stefan.scherer,friedhelm.schwenker,
guenther.palm}@uni-ulm.de

Nick Campbell
Center for Language and Communication Studies, Trinity College Dublin
e-mail: nick@tcd.ie

H. Ritter et al. (Eds.): Human Centered Robot Systems, COSMOS 6, pp. 111–120.
springerlink.com

communication modalities, e.g. gestures etc., to communicate with human dialog partners [15]. Furthermore, laughter is acoustically highly variable and is expressed in many forms, such as giggles, exhaled or inhaled laughs, or even snort like laughters exist. Therefore, it is suspected, that laughter is difficult to model and to detect [1, 17].

However, modeling laughter and thereby detecting laughter in natural discourses has been the topic of related research: in [8] one second large segments of speech are considered. For each of these segments the decision, whether somebody of the speakers laughed or not, is being made using Mel Frequency Cepstral Coefficients (MFCC) and Support Vector Machines (SVM). The recognition accuracy of this approach reached 87%. One of the obvious disadvantages of this approach is that segments of one second in length are used and therefore no accurate on- and offsets of the laughs can be detected.

Truong and Leeuwen [16, 17] first recognized laughter in previously segmented speech data taken from a manually labeled meeting corpus containing data from close head mounted microphones. They used Gaussian Mixture Models (GMM) and pitch, modulation spectrum, perceptual linear prediction (PLP) and energy related features. They achieved the best results of 13.4% equal error rate (EER) using PLP features on pre-segmented audio samples of an average length of 2 seconds for laughter and 2.2 seconds for speech. In their second approach [17] they extracted PLP features from 32 ms windows every 16 ms using three GMMs for modeling laughter, speech, and silence. Silence was included since it was a major part of the meeting data. An EER on segmenting the whole meeting of 10.9% was achieved. In future work they want to use HMMs to further improve their results. Their second approach allows a very accurate detection of laughter on- and offsets every 16 ms. However, it does not consider the, in [9] mentioned, average length of a laughter segment of around 200 ms since only 32 ms of speech are considered for each decision.

In [9] the same data set as in [16, 17] was used for laughter detection. In a final approach after narrowing down the sample rate of their feature extractor to a frequency of 100 Hz (a frame every 10 ms) a standard Multi Layer Perceptron (MLP) with one hidden layer was used. The input to the MLP was updated every 10 ms, however the input feature vector considered 750 ms including the 37 preceding and following frames of the current frame for which the decision is computed by the MLP. The extracted features include MFCCs and PLPs since they are perceptually scaled and were chosen in previous work. Using this approach an EER of around 7.9% was achieved.

In the current work Echo State Networks (ESN) are used to recognize laughter in a meeting corpus [2], comprising audio and video data for multimodal detection experiments. The approach utilizing ESNs, is making use of the sequential characteristics of the modulation spectrum features extracted from the audio data [7]. Furthermore, the features are extracted every 20 ms and comprise data of 200 ms in order to be able to give accurate on- and offset positions of laughter, but also to comprise around a whole "laughter syllable" in one frame [9]. In a second approach the video data containing primary features such as head and body movement are incorporated into the ESN approach for a multimodal laughter detection method. One of the main goals of this work is to provide a classification approach that is only relying on unobtrusive recording gear, such as a centrally placed microphone and a 360 degrees camera, since it is particularly important to provide a natural environment for unbiased communication. However, this constraint only allows the extraction and analysis of basic features.

The remainder of this paper is organized as follows: Section 2 gives an introduction on the used data and explains the recording situation in detail, Section 3 describes the utilized features and the extraction algorithm, Section 4 comprises detailed descriptions of the approaches for classifying the data. Section 5 reports the obtained results of the single and multimodal approaches. Finally, Section 6 concludes the paper and summarizes the work.

2 Utilized Data

The data for this study consists of three 90 minutes multi-party conversations in which the participants originating from four different countries each speaking a different native language. However, the conversation was held in English. The conversations were not constrained by any plot or goal and the participants were allowed to move freely, which renders this data set very natural and therefore it is believed that the ecological validity of the data is very high. The meetings were recorded by using centrally positioned, unobtrusive audio and video recording devices. The audio stream was directly used as input for the feature extraction algorithm at a sample rate of 16 kHz. A 360 degree video capturing device was used for video recording and the standard Viola Jones algorithm was used to detect and track the faces of the participants throughout the 90 minutes. The resulting data has a sample rate of 10 Hz and comprises head and body activity [2]. In Figure 1 one of the 360 degree camera frames including face detection margins is seen. It is clear that only the heads and upper parts of the body are visible since the participants are seated around a table. However, hand gestures, head and body movements are recorded without any obstruction. Separate analysis has revealed high correlations of activity between body and head movement of active speakers, as well as it is expected that the coordinates of the head positions should move on a horizontal axis while the speaker produces a laughter burst, which could be a sequence learnt by the ESN.

The little constraints on the conversation provide very natural data including laughters, and other essential paralinguistic contents. The data was annotated manually and non-speech sounds, such as laughters or coughs were labeled using symbols indicating their type. Laughter including speech, such as a laughter at the end of an utterance, was labeled accordingly, but was not used as training data for the classifiers in order not to bias the models by the included speech. However, all the data was used in testing. For training a set of around 300 laughters containing no speech of an average length of 1.5 seconds and around 1000 speech samples of an average length of 2 seconds are used. A tenth of this pool of data was excluded from training

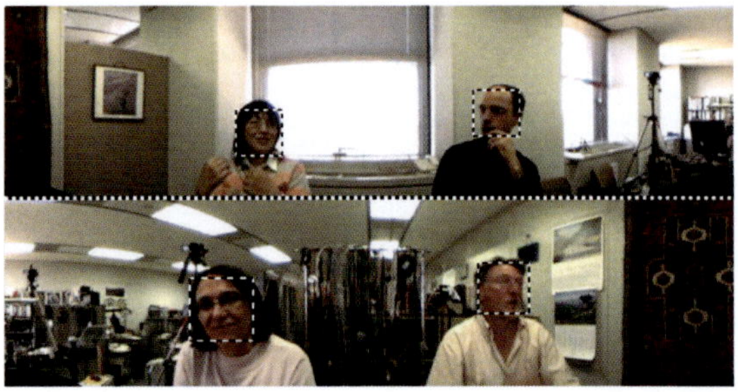

Fig. 1 A frame of the 360 degree camera positioned in the center of the conference table. The image includes the boundaries of the detected faces using the Viola Jones approach.

in each fold of the 10-fold cross validations, except in the experiments in which all the dialog data is presented, including all the laughs including speech in the labeled segment. Overall, laughter is present in about 5-8% of the data. This variance is due to the laughters including speech[1].

3 Features

As mentioned before the available data for the experiments is comprised of two different modalities. The conversations are available as audio and video files. Therefore, suitable features for the laughter detection task were extracted. In the following two paragraphs the extraction procedures are explained in some detail, for further information refer to the cited articles.

For the detection of laughter we extracted modulation spectrum features from the audio source [5]. These features have been used in previous experiments and tasks such as emotion recognition and laughter recognition [13, 16, 11]. The features are extracted using standard methods like Mel filtering and Fast Fourier Transformations (FFT). In short they represent the rate of change of frequency, since they are based on a two level Fourier transformation. Furthermore, they are biologically inspired since the data is split up into perceptually scaled bands. In our experiment we used 8 bands in the Mel filtering regarding frequencies up to 8 kHz. Since the audio is sampled at a rate of 16 kHz this satisfies the Nyquist theorem [5]. These slow temporal modulations of speech emulate the perception ability of the human auditory system. Earlier studies reported that the modulation frequency components from the range between 2 and 16 Hz, with dominant component at around 4 Hz, contain important linguistic information [4, 3]. Dominant components represent strong rate of change of the vocal tract shape.

[1] The data, and annotations are freely available on the web, but access requires a password and user name, which can be obtained from the authors on request, subject to conditions of confidentiality.

As mentioned before the video features were extracted from the 360 degree recordings of a centrally placed camera. The sample rate of the face tracking using the Viola Jones approach was 10 Hz and the provided data comprised coordinates of the faces at each frame composed by the exact spot of the top left corner and the bottom right corner of the surrounding box of the face as seen in Figure 1. However, these coordinates are highly dependent on the distance of the person to the camera and therefore relative movement data of the face and body were taken as input to the classifiers. The coordinates were normalized to movement data of a mean value of 0 and a standard deviation of 1. Therefore, the movement ranged from -1 to 1 for each tracked face and body individually.

4 Echo State Network Approach

For the experiments a relatively novel kind of recurrent neural networks (RNN) is used, the so called Echo state network (ESN) [7]. Among the advantages of an ESN over common RNNs are the stability towards noisy inputs [14] and the efficient method to adapt the weights of the network [6]. Using the direct pseudo inverse adaptation method the ESN is trained in a very efficient way. With regard to these advantages and considering the targeted application area of the network the ESN is a fitting candidate. In contrast to for example SVMs used in [8] for the detection of laughter the ESN incorporates previous features and states for the decision whether or not a laughter is present, rendering it an ideal approach for online detection tasks.

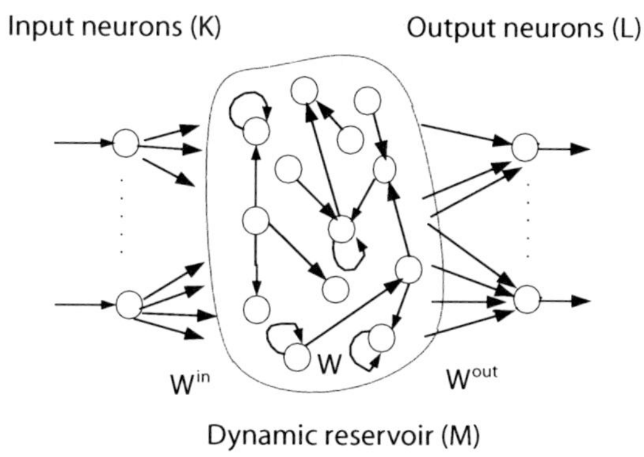

Fig. 2 Scheme of an Echo state network

In Figure 2 the scheme of an ESN is shown. The input layer K is fully connected to the dynamic reservoir M, and M is fully connected to the output layer L. ESNs are characterized by their dynamic memory that is realized by a sparsely interconnected reservoir M that is initialized randomly. The connection matrix is normalized to a so called spectral width parameter α guaranteeing that the activity within the dynamic reservoir is kept at a certain level. In order to train an ESN it is only necessary to adapt the output weights W^{out} using the direct pseudo inverse method computing the optimal values for the weights from M to L by solving the linear equation system $W^{out} = M^{+}T$. The method minimizes the distance between the predicted output of the ESN and the target signal T. For the detailed algorithm refer to [6].

After training the ESN output is being post processed in order to avoid rapid shifts between states. Stability is being achieved by several smoothing steps as follows: First the output is smoothed using a fourth grade Butterworth filter with a cut off frequency rate of 0.3. If the output exceeds a threshold $\theta = 0.4$ it is counted as a hit and the output is set to 1, values below are set to 0. After generating this binary vector a Gaussian filter of a length of 25 is applied in order to get the outputs shown in the figures in the following section.

5 Experiments and Results

The basis architecture used are the ESNs introduced in Section 4. The ESN consist of a dynamic reservoir with 1500 neurons that are sparsely interconnected with each other. The probability for a connection between neuron x and y is 2%. Recursive connections are also set at the same probability. The last parameter of the ESN that has to be set is the spectral width influencing the dynamics of the reservoir. The spectral width α was set to 0.15.

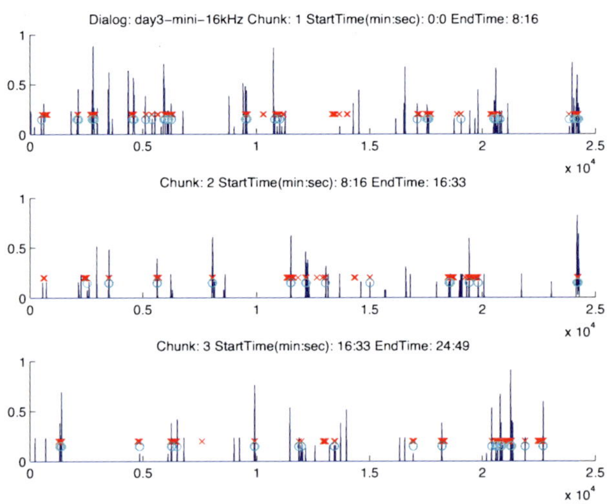

Fig. 3 Fusion Echo state network output after post processing. The blue line corresponds to the output, blue circles depict hits and red crosses correspond to human labels.

In a first experiment a 10-fold cross validation was conducted on the speech data and laughter comprising no speech. The ESN recognizes laughter on each frame provided by the feature extraction. The knowledge of on- and offsets of utterances or laughter provided by the labels is not utilized for the classification. An average error rate of around 13% was achieved by the ESN.

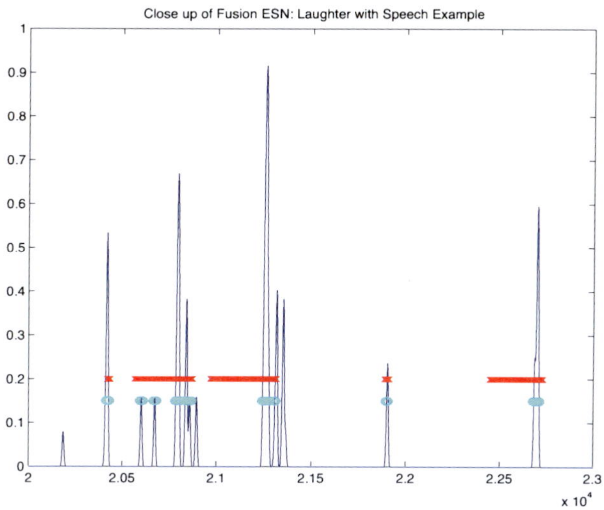

Fig. 4 Close up look to the Echo state network output after post processing. The blue line corresponds to the output, blue circles depict hits and red crosses correspond to human labels.

In a second series of a 10-fold cross validation the whole dialog is given as input to the ESN. The classification accuracy was again around 90%. An average misclassification rate of 10.5% was achieved over the 10 folds. The increase in accuracy can be explained as a result of overlapping training and test data. However, these percentages are biased as already mentioned before, since the ESN was only trained on laughter containing no speech and the whole dialog contains laughters that are labeled together with speech in some cases. Therefore, a more subjective view on the results is necessary. In Figure 3, the first three parts out of ten of the conversation are seen. Laughter labels are indicated by red crosses and the output of the ESN after post-processing as described in Section 4 is displayed in blue. The ESN clearly peaks most of the time at the labeled laughters. Only a few laughters are omitted and some are inserted. Furthermore, it is interesting that some labels are quite long, but only at the end or beginning the ESN peaks as in Figure 4 at around 22000 on the x axis. In this particular case the laughter in the conversation appears at the end of the utterance labeled by the human labelers. Therefore, the system could be used for post-processing of the manual labels and refine them.

In a the final experiment we made use of the available video data in order to test the networks performance to detect laughter in a multimodal approach. The available video data only comprises basic features such as head and body movement for each speaker and x and y coordinate changes of the head frame by frame. As input we made use of the body and head movement and normalized them for each speaker towards an average of 0 and variance 1. Using this approach we receive 8 dimensional features at a sampling rate of 10 Hz due to the output of the Viola Jones face recognition algorithm. In order to be able to use the same ESN architecture for the movement data we had to adapt the sampling rate of the data. This is done by simply memorizing the movement ESN output for 5 frames instead of only one. The final architecture incorporates the output of two separate ESNs in the two modalities in a weighted sum before post processing steps are taken. After training in the test phase the outputs are added using different weighting for each ESN. Thorough testing and considering the coarseness of the video data and the related bias, resulted in a weighting of 0.7 for the audio related ESN and a weight of 0.3 for the movement ESN. This fusion is then post-processed as the audio ESN output in the first two experiments. Using this fusion we obtain less false alarms and less misses. Therefore, the overall performance got better. However, the result might not be significantly better since the improvement only resulted in an error of 9%. Further, numerical studies need to be carried out to test statistical significance. In Figure 5 a comparison of the three ESN modalities speech, movement, and fusion is given. It is seen that the fusion output is less unstable and calmer in comparison to the other two. The output for the movement data is the most unstable output and resulted in around 18% error.

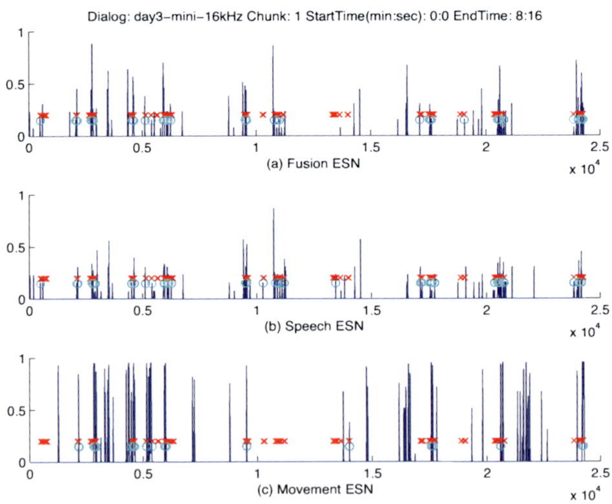

Fig. 5 Comparison between the three possible modalities: (a) fusion approach, (b) speech only and (c) movement only

6 Summary and Discussion

The work presented in this paper aims towards the online detection of laughter in speech. Multi-party natural discourse data was used as data, consisting of three videos of around 90 minutes in length. The conversations were not restricted by any constraints and were therefore as natural as possible, comprising all sorts of gestures, laughters, and noise sources such as drinking a cup of coffee or people entering the room. This naturalness of the conversation was supported further by unobtrusive recording devices placed centrally on the table. The participants themselves were not wired with any close up microphone nor intimidated by any camera facing them.

This multimodal data was then used in several recognition experiments. Results comparable to related work were achieved using multimodal data and ESNs. On the other hand the ESN takes the extracted features covering 200 ms of audio corresponding to the average length of a laughter syllable [9] directly as input. In a series of experiments in which the task was to detect laughter in a continuous stream of input the ESN compensates the lack of information by memorizing previous decisions and features and utilizing this information in order to predict upcoming events. The ESN approach single- and multimodal can be used as a tool for the online recognition of laughter. The detected laughter may then be used to directly control robots in human robot interaction scenarios. Since, a robot that may react appropriately in a timely fashion to laughs or smiles seems more natural than a robot that is incapable of reacting to such events in real time [12].

For future work we aim towards a method to measure "engagement" in discourse, which can be applied to measure the quality of interaction between humans as well as between a human and an artificial agent. Previously recorded data, such as in [15], will be labeled accordingly and an automatic recognition approach using multimodal data as input to ESN ensembles will be constructed. Since laughter is an indicator for the positive perception of a discourse element, we consider the detection of laughter an essential part aiming for the goal of measuring the quality of interaction, and the degree of interaction, as laughter is one of the main features to detect participant involvement [10]. An agent that keeps you engaged in a conversation will be perceived more positive than one that bores you.

Acknowledgements. This research would not have been possible without the generous funding and the kind support of the German Academic Exchange Agency (DAAD). Furthermore, the presented work was developed within the Transregional Collaborative Research Centre SFB/TRR 62 "Companion-Technology for Cognitive Technical Systems" funded by the German Research Foundation (DFG).

References

1. Campbell, N., Kashioka, H., Ohara, R.: No laughing matter. In: Proceedings of Interspeech, ISCA, pp. 465–468 (2005)
2. Campbell, W.N.: Tools and resources for visualising conversational-speech interaction. In: Proceedings of the Sixth International Language Resources and Evaluation (LREC 2008), ELRA, Marrakech, Morocco (2008)

3. Drullman, R., Festen, J., Plomp, R.: Effect of reducing slow temporal modulations on speech reception. Journal of the Acousic Society 95, 2670–2680 (1994)
4. Hermansky, H.: Auditory modeling in automatic recognition of speech. In: Proceedings of Keele Workshop (1996)
5. Hermansky, H.: The modulation spectrum in automatic recognition of speech. In: Proceedings of IEEE Workshop on Automatic Speech Recognition and Understanding, pp. 140–147. IEEE, Los Alamitos (1997)
6. Jaeger, H.: Tutorial on training recurrent neural networks, covering bppt, rtrl, ekf and the echo state network approach. Tech. Rep. 159, Fraunhofer-Gesellschaft, St. Augustin Germany (2002)
7. Jaeger, H., Haas, H.: Harnessing nonlinearity: Predicting chaotic systems and saving energy in wireless communication. Science 304, 78–80 (2004)
8. Kennedy, L., Ellis, D.: Laughter detection in meetings. In: Proceedings of NIST ICASSP, Meeting Recognition Workshop (2004)
9. Knox, M., Mirghafori, N.: Automatic laughter detection using neural networks. In: Proceedings of Interspeech 2007, ISCA, pp. 2973–2976 (2007)
10. Laskowski, K.: Modeling vocal interaction for text-independent detection of involvement hotspots in multi-party meetings. In: Proceedings of the 2nd IEEE/ISCA/ACL Workshop on Spoken Language Technology (SLT 2008), pp. 81–84 (2008)
11. Maganti, H.K., Scherer, S., Palm, G.: A novel feature for emotion recognition in voice based applications. In: Paiva, A.C.R., Prada, R., Picard, R.W. (eds.) ACII 2007. LNCS, vol. 4738, pp. 710–711. Springer, Heidelberg (2007)
12. Pugh, S.D.: Service with a smile: Emotional contagion in the service encounter. Academy of Management Journal 44, 1018–1027 (2001)
13. Scherer, S., Hofmann, H., Lampmann, M., Pfeil, M., Rhinow, S., Schwenker, F., Palm, G.: Emotion recognition from speech: Stress experiment. In: Proceedings of the Sixth International Language Resources and Evaluation (LREC 2008). European Language Resources Association (ELRA), Marrakech, Morocco (2008)
14. Scherer, S., Oubbati, M., Schwenker, F., Palm, G.: Real-time emotion recognition from speech using echo state networks. In: Prevost, L., Marinai, S., Schwenker, F. (eds.) ANNPR 2008. LNCS (LNAI), vol. 5064, pp. 205–216. Springer, Heidelberg (2008)
15. Strauss, P.M., Hoffmann, H., Scherer, S.: Evaluation and user acceptance of a dialogue system using wizard-of-oz recordings. In: 3rd IET International Conference on Intelligent Environments, IET, pp. 521–524 (2007)
16. Truong, K.P., Van Leeuwen, D.A.: Automatic detection of laughter. In: Proceedings of Interspeech, ISCA, pp. 485–488 (2005)
17. Truong, K.P., Van Leeuwen, D.A.: Evaluating laughter segmentation in meetings with acoustic and acoustic-phonetic features. In: Workshop on the Phonetics of Laughter, Saarbrücken, pp. 49–53 (2007)

Classifier Fusion Applied to Facial Expression Recognition: An Experimental Comparison

Martin Schels, Christian Thiel, Friedhelm Schwenker, and Günther Palm

Abstract. In this paper classifier fusion approaches are investigated through numerical evaluation. For this purpose a multi classifier architecture for the recognition of human facial expressions in image sequences has been constructed on characteristic facial regions and three different feature types (principal components, orientation histograms of static images and temporal features based on optical flow). Classifier fusion is applied to the individual channels established by feature principle and facial region, which are addressed to by individual classifiers. The available combinations of classifier outputs are examined and it is investigated how combining classifiers can lead to more appropriate results. The stability of fusion regarding varying classifier combinations is studied and the fused classifier output is compared to the human view on the data.

1 Introduction and Related Work

Combining classifiers via classifier fusion is a well established technique in pattern recognition. But in spite of theoretical studies it is not yet known how to best construct a combination of classifiers in general, thus there is still a need of empirical evaluation for every particular application.

In [12] static and trainable classifier fusion approaches were evaluated regarding to a multi-modal person verification application. For this, six individual classifiers were constructed on video and audio data, and experiments were conducted in order to investigate the impact on accuracy of a rising number of classifiers contributing to a fused classifier and the effects of fusing classifiers with imbalanced performances. It was affirmed that combining complementary classifiers is more effective than weighting classifiers only by their individual performance. An improvement

Martin Schels · Christian Thiel · Friedhelm Schwenker · Günther Palm
Institute of Neural Information Processing, University of Ulm, 89069 Ulm, Germany
e-mail: {martin.schels,christian.thiel,friedhelm.schwenker, guenther.palm}@uni-ulm.de

H. Ritter et al. (Eds.): Human Centered Robot Systems, COSMOS 6, pp. 121–129.
springerlink.com

of the classification accuracy was achieved using fusion, but there was no relation found between an increasing number of experts and a decrease of the error rate. Furthermore, it was found that static fusion approaches do perform well when the contributing classifiers perform approximately even, and trainable fusion was able to incorporate bigger differences of performance. But this behavior is found to be strongly tied to the quantity and quality of training data.

In [7] several so-called fixed fusion rules were evaluated, applied to a hand written digit recognition task. Four different individual classifiers were trained, each using a different classifier principle. In this study it is was also emphasized, that the single classifiers' outputs should produce independent errors to be suitable for classifier fusion. This could be achieved in this study by using different feature views on the data or utilizing various classifier principles, which make different assumptions about the underlying data. Thus the recognition performance could be improved with respect to the best single classifier in several cases, especially for this application the sum rule performed well, which was more deeply examined.

In [8] another experimental evaluation of classifier fusion can be found. The main scope of this work was the evaluation of Decision Templates, but plenty other approaches were also addressed. The experiments were conducted on the Satimage and the Phoneme data set, which are part of the ELENA collection, and six respectively ten individual classifiers were constructed. There was only little improvement of the combined classifier over the best single classifier and the authors propose that increasing the independence of the different training sets could lead to a greater benefit from the classifier fusion. But an interesting result of their study is that simple static fusion rules do well, but not with every data. And even though they report Decision Templates to be superior in their experiments, they state that there is no fixed approach, which is to be superior at any application.

The detection and recognition of the user's emotional state is an important aspect for the design of more natural interfaces in human computer interaction applications. In face-to-face interaction of humans hand gestures and facial expressions convey nonverbal communication cues and therefore the automatic recognition of such expressions can also play an important role in human computer interaction. In this paper we focus the recognition of facial expressions. There is a rich body of literature available on the detection of emotions in human facial expressions, see [4] for a survey.

The main objective of our study is to investigate and to design ensembles of classifiers and to study the problem of classifier fusion in this application. Several fusion approaches are numerically evaluated and the effects of combining different classifiers with a varying performances are investigated.

2 Classifiers and Classifier Fusion

In the following we give a brief introduction to the SVM and RBF-network classifiers we use, followed by an explanation of the different approaches for classifier fusion. Basically, Support Vector Machines (SVM) are binary classifiers that can be extended to multi-class SVMs (see [14] for an introduction on SVM). In our study SVMs implementing the one-against-one decomposition, which is extended to fuzzy-input fuzzy-output classification are utilized [17].

RBF-Networks are multilayer networks [3, 10], which have distance computing neurons with a radial basis transfer function in the hidden layer. The parameters of RBF networks were trained by gradient descent using the RPROP technique [11].

Combining multiple classifiers can produce a more accurate output than a single one. As mentioned in the introduction an important constraint for an improvement of the recognition rate is that the ensemble of classifiers should produce independent errors. This can be achieved by using different feature views of a dataset or by using different classifier architectures or by utilizing different subsamples of the training data set.

Classifier fusion approaches, that implement a fixed rule without any training procedure are called static. In our experiments minimum, maximum, mean and product rule fusion were evaluated. These fusion rules apply the minimum, maximum, mean or product operator within the particular classes.

Decision Templates and pseudoinverse solution are examples for trainable fusion approaches [15]. Both mappings are using the following form, implementing different transformation matrices V^i, which is incorporating the confusion matrix YC_i^T of the individual classifiers: $z = \sum_{i=1}^{N} V^i C^i(x)^T$. Here z denotes the fused output and $C^i(x)$ is the output of the i-th classifier for data-vector x. The variable Y denotes the desired classifier output. For Decision Templates [15] the matrix is specified as $V^i = (YY^T)^{-1}(YC_i^T)$. The multiplication of $(YY^T)^{-1}$ with the confusion matrix normalizes it with the number of samples of a class. This formulation of Decision Templates is equivalent calculating to the means of the classifiers for each class. Pseudoinverse [15] solution calculates a least-squares linear mapping from the output of the classifiers to the desired output. The matrix V_i is calculated as $V^i = Y \lim_{\alpha \to 0_+} C_i^T (C_i C_i^T + \alpha I)^{-1}$. The limit ensures that the covariance matrix is always invertible.

We also evaluated a brute force fusion approach: Another SVM is constructed as fusion layer, which employs inputs the concatenated outputs of the individual classifiers as input. Do note, that this mapping is not restricted to a linear mapping as the trainable fusion approaches described above.

3 Data Collection

The Cohn-Kanade dataset is a collection of image sequences with emotional content [6], which is available for research purposes. It contains image sequences, which were recorded in a resolution of 640×480 (sometimes 490) pixels with a temporal resolution of 33 frames per second. Every sequence is played by an amateur actor who is filmed from a frontal view. The sequences always start with a neutral facial

Table 1 Confusion matrix of the human test persons against the majority of all 15 votes (left). The right column shows the share of the facial expressions in the data set (hardened class labels).

maj.\test pers.	hap.	ang.	sur.	disg.	sad.	fear	no. samples
hap.	**0.99**	0	0	0	0	0.01	105
ang.	0	**0.8**	0	0.12	0.07	0.01	49
sur.	0.01	0	**0.78**	0	0.01	0.19	91
disg.	0.01	0.15	0.01	**0.67**	0.01	0.15	81
sad.	0	0.08	0.02	0.02	**0.88**	0.01	81
fear	0.01	0.01	0.14	0.27	0.01	**0.56**	25

expression and end with the full blown emotion which is one of the six categories "fear", "happiness", "sadness", "disgust", "surprise" or "anger".

To acquire a suitable label the sequences were presented to 15 human labelers (13 male and two female). The sequences were presented as a video. After the playback of a video the last image remained on the screen and the test person was asked to select a label. Thus, a fuzzy label for every sequence was created as the mean of the 15 different opinions. The result of the labeling procedure is given in Tab. 1, showing the confusion matrix of the test persons according to the majority of all persons. It is revealed that the data collection is highly imbalanced only 25 samples expression "fear" occur in the data set and in addition, this expression could not be identified by the test persons.

In all automatic facial expression recognition systems first some relevant features are extracted from the facial image and these feature vectors then utilized to train some type of classifier to recognize the facial expression. One problem is here how to categorize the emotions: one way is to model emotions through a finite set of emotional classes such as anger, joy, sadness, etc, another way is to model emotions by a continuous scales, such as valence (the pleasantness of the emotion) and arousal (the level of activity) of an expression [9]. In this paper we use a discrete representation in six emotions. Finding the most relevant features the definitely the most important step in designing a recognition systems. In our approach prominent facial regions such as the eyes, including the eyebrows, the mouth and for comparison the full facial region have been considered. For these four regions orientation histograms, principal components, optical flow features have been computed. Principal components (eigenfaces approach) are very well know in face recognition [18], and orientation histograms were successfully applied for the recognition of hand gestures [5] and faces [16], both on single images. In order to extract the facial motion in these regions optical flow[1] features from pairs of consecutive images have been computed, as suggested in [13].

[1] We were using a biologically inspired optical flow estimator, which was developed by the Vision and Perception Science Lab of the Institute of Neural Processing at the University of Ulm [2, 1] .

4 Experiments and Results

The classification experiments are all executed in the following set-up: For every distinct combination of feature and region a SVM or a RBF-network operating on a single frame, i.e. feature of an image or pair of successive images is trained. The classifiers are optimized using 8-fold cross validation and all the following results are also 8-fold cross validated. A list of classifiers and the individual performances for each of these channels can be found in Tab. 2, evidently our classifiers show rather imbalanced recognition rates concerned to the emotional categories. The frame-wise results are temporally integrated by taking the average over the decisions [2]. These results are then fused into a global result using one of the approaches presented in Sect. 2.

We evaluate the whole combinatory space, which is given by the 14 available classifiers, for every fusion approach to study its properties on different feature combinations. The best results of the fusion steps are noted in Tab. 3, which shows

Table 2 Overview of the individual classifiers ordered by recognition rate. These 14 single classifiers are the candidates for our classifier fusion experiments.

No.	Feature	Region	Classifier	Rec.-Rate (%)
1	Orientation histograms	mouth	SVM	74.0
2	Orientation histograms	mouth	RBF	70.3
3	Optical flow	face	RBF	67.1
4	Optical flow	mouth	RBF	67.1
5	PCA	mouth	RBF	65.9
6	PCA	face	RBF	62.7
7	Orientation histograms	face	SVM	54.3
8	Orientation histograms	face	RBF	52.3
9	Optical flow	right eye	RBF	51.1
10	Optical flow	left eye	RBF	45.8
11	Orientation histograms	right eye	RBF	42.3
12	Orientation histograms	left eye	RBF	41.4
13	PCA	left eye	RBF	37.9
14	PCA	right eye	RBF	35.1

also an entry called "oracle" [8]. This is the amount of samples, that is correctly classified by any classifier. It is of course not one of the proposed approaches for classifier fusion, but it should give a hint to the potential of the ensemble. With a value of 98.3 % this measure is quite promising in our case. Comparing the fused results with the best single classifier shows that nearly every fusion approach leads to an improvement of the recognition rate. Only the usage of maximum fusion has a

[2] It was also considered to to use Hidden Markov Models for this step, which did not turn out to be feasible.

Table 3 Fusion results of experiments with static and trainable fusion. Also listed are the best single machine classifier (compare Tab. 2) and the average performance of the human labelers against the mean of all labelers and the "oracle" [8].

Fusion approach	Rec.-rate (%)
Product rule	83.1
Mean	81.7
Minimum	77.8
Maximum	73.8
Brute force (SVM)	83.6
Pseudoinverse	81.4
Decision-Templates	75.9
Best single classifier	74.0
Average human labeler	81.5
"Oracle"	98.3

negative effect. Applying brute force as fusion method results in a recognition rate of 83.6 %, which is the highest rate in this study. This even slightly outperforms the average human test person (81.5 %), as determined in the labeling experiments. Many static fusion approaches, namely mean fusion and product rule, result in high recognition rates. Only one of the classical trainable fusion approaches, which are mainly linear mappings, does only result an acceptable recognition rate: pseudoinverse fusion is able to reach up to 81.4 % correctly classified samples, while decision tables can only slightly improve over the best individual classifier.

To examine the performance on varying combinations, we are sorting the results by the percentage of correctly classified samples and classifier combination within the fusion approaches (see Fig. 1). It can be observed that the pseudoinverse approach outperforms the best individual classifier more often than probabilistic product and brute force, which are fusion approaches with higher recognition rates in Tab. 3. So the training of the pseudoinverse mapping does result in more robust results and it should be easier to pick a classifier combination for an application. This observation sustains for this case the claim, that trainable fusion could incorporate weaker classifiers better [12] because of the training procedure. On the other hand the brute force fusion approach does result in the highest over-all recognition rate, but proves to be even more unstable than the simple product rule.

In Tab. 4 (bottom) the confusion matrix of the champion architecture using the product rule for the classifiers 2, 4, 6, 9 and 10 is displayed. The fused classifier reveals a quite similar recognition performance as the human test persons (see Tab. 1), referring to the particular classes. All the classes except "surprise" and "fear" show an almost identical performance on the diagonal line of the confusion matrices in both cases. For the machine classifier the task seems to imply similar difficulties as for the test persons. It is obvious that our classifier performs only weak for the class "fear", similarities can be observed: In both cases this class is confused with classes "surprise" and "disgust" more often than others.

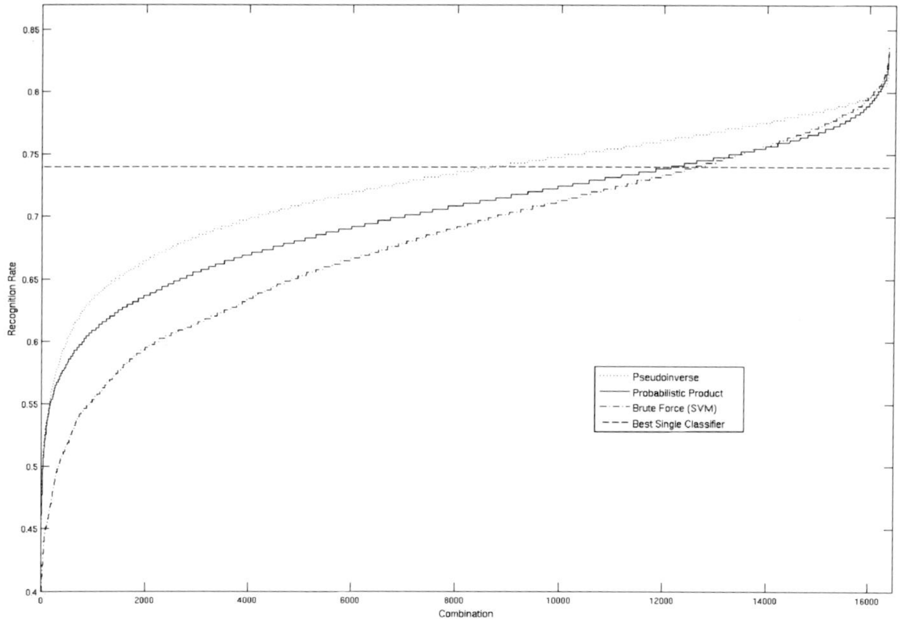

Fig. 1 Sorted recognition rates of three selected fusion approaches. Pseudoinverse fusion does not produce the highest recognition rates of the evaluated approaches, but this mapping does outperform the best individual classifier more often.

Table 4 Top: Confusion matrix of two classifiers (left: classifier 4; right: classifier 6), which are contributing to the best classifier architecture (bottom) using probabilistic product (built out of classifiers 2, 4, 6, 9 and 10). The entries of the matrices are fractions.

true\classif.	hap.	ang.	sur.	disg.	sad.	fear
happiness	**0.87**	0.1	0.1	0.8	0.3	0.1
anger	0.2	**0.59**	0.2	0.4	0.29	0.0
surprise	0.1	0.0	**0.89**	0.4	0.2	0.3
disgust	0.25	0.20	0.7	**0.32**	0.10	0.6
sadness	0.7	0.20	0.1	0.5	**0.67**	0.0
fear	0.28	0.4	0.28	0.20	0.8	**0.12**

true\classif.	hap.	ang.	sur.	disg.	sad.	fear
happiness	**0.79**	0.2	0.4	0.8	0.4	0.4
anger	0.12	**0.53**	0.0	0.12	0.22	0.0
surprise	0.9	0.1	**0.82**	0.3	0.3	0.1
disgust	0.20	0.19	0.2	**0.37**	0.16	0.6
sadness	0.10	0.15	0.5	0.5	**0.64**	0.1
fear	0.12	0.0	0.20	0.28	0.16	**0.24**

true\classif.	hap.	ang.	sur.	disg.	sad.	fear
happiness	**0.95**	0.01	0	0.04	0	0
anger	0	**0.80**	0	0.10	0.10	0
surprise	0.01	0	**0.97**	0	0.01	0.01
disgust	0.17	0.12	0.01	**0.63**	0.01	0.05
sadness	0.01	0.05	0.04	0.01	**0.89**	0
fear	0.08	0	0.20	0.32	0.04	**0.36**

Table 4 shows also two confusion matrices of classifiers 4 and 6, which are part of the ensemble forming the third confusion matrix. Now we can exemplarily investigate the possible impact of the fusion procedure: Classifier 4 is found to be superior to classifier 6 in class "happiness" and "surprise", but inferior in classes "disgust" and especially "fear". In the final fused classifier these results are merged and in all cases there is a further improvement in all classes.

5 Summary and Conclusion

In this paper we studied various multiple classifier fusion approaches applied to the classification of human facial expressions. To this end we trained individual classifiers for characteristic facial segments (left and right eye, mouth and full face) and three feature types. Trainable and static fusion approaches were examined concerning the incorporation of classifiers with different performances. The linear transformation fusion with pseudoinverse shows at this application greater stability with respect to the combination of individual classifiers and does also result in high recognition rates. Brute force and the static fusion rules did provide less stability even though some of these techniques do deliver a high top recognition rate.

To motivate classifier fusion, the behavior of two individual classifiers was exemplarily inspected by analyzing the confusion matrices. The classifiers supplemented each other by producing different errors and thus the fused classifier is able to reach a better performance. By comparing the confusion matrix of a fused ensemble to the confusion matrix of the human test persons a remarkable analogies were observed.

Acknowledgements. The authors would like to thank Prof. Dr. H. Neumann, Dr. P. Bayerl and S. Ringbauer from the Vision and Perception Science Lab of the Institute of Neural Processing at the University of Ulm for generous assistance in extracting the optical flow features.

This paper is based on work done within the project SCHW623/4-3, and the "Information Fusion" subproject of the Transregional Collaborative Research Centre SFB/TRR 62 "Companion-Technology for Cognitive Technical Systems", both funded by the German Research Foundation (DFG). The work of Martin Schels is supported by a scholarship of the Carl-Zeiss Foundation.

References

1. Bayerl, P., Neumann, H.: Disambiguating visual motion through contextual feedback modulation. Neural Comput. 16, 2041–2066 (2004)
2. Bayerl, P., Neumann, H.: A fast biologically inspired algorithm for recurrent motion estimation. IEEE Transactions on Pattern Analysis and Machine Intelligence 29, 246–260 (2007)
3. Bishop, C.M.: Pattern Recognition and Machine Learning. Springer, Heidelberg (2006)

4. Cowie, R., Douglas-Cowie, E., Tsapatsoulis, N., Votsis, G., Kollias, S., Fellenz, W., Taylor, J.: Emotion recognition in human-computer interaction. IEEE Signal Processing Magazine 18, 32–80 (2001)
5. Freeman, W.T., Roth, M.: Orientation Histograms for Hand Gesture Recognition. In: International Workshop on Automatic Face and Gesture Recognition, pp. 296–301 (1994)
6. Kanade, T., Cohn, J., Tian, Y.L.: Comprehensive database for facial expression analysis. In: Proceedings of the 4th IEEE International Conference on Automatic Face and Gesture Recognition (FG 2000), pp. 46–53 (2000)
7. Kittler, J., Hatef, M., Duin, R.P.W., Matas, J.: On combining classifiers. IEEE Trans. Pattern Anal. Mach. Intell. 20, 226–239 (1998)
8. Kuncheva, L., Bezdek, J.C., Duin, R.P.W.: Decision templates for multiple classifier fusion: an experimental comparison. Pattern Recognition 34, 299–314 (2001)
9. Lang, P.J.: The emotion probe. studies of motivation and attention. The American Psychologist 50, 372–385 (1995)
10. Poggio, T., Girosi, F.: A theory of networks for approximation and learning. Laboratory, Massachusetts Institute of Technology 1140 (1989)
11. Riedmiller, M., Braun, H.: RPROP – description and implementation details. Technical report, Universität Karlsruhe (1994)
12. Roli, F., Kittler, J., Fumera, G., Muntoni, D.: An experimental comparison of classifier fusion rules for multimodal personal identity verification systems. In: Roli, F., Kittler, J. (eds.) MCS 2002. LNCS, vol. 2364, pp. 325–336. Springer, Heidelberg (2002)
13. Rosenblum, M., Yacoob, Y., Davis, L.: Human expression recognition from motion using a radial basis function network architecture. IEEE Transactions on Neural Networks 7, 1121–1138 (1996)
14. Schölkopf, B., Smola, A.J.: Learning with Kernels: Support Vector Machines, Regularization, Optimization, and Beyond (Adaptive Computation and Machine Learning). The MIT Press, Cambridge (2001)
15. Schwenker, F., Dietrich, C., Thiel, C., Palm, G.: Learning of decision fusion mappings for pattern recognition. International Journal on Artificial Intelligence and Machine Learning (AIML) 6, 17–21 (2006)
16. Schwenker, F., Sachs, A., Palm, G., Kestler, H.A.: Orientation Histograms for Face Recognition. In: Schwenker, F., Marinai, S. (eds.) ANNPR 2006. LNCS (LNAI), vol. 4087, pp. 253–259. Springer, Heidelberg (2006)
17. Thiel, C., Giacco, F., Schwenker, F., Palm, G.: Comparison of Neural Classification Algorithms Applied to Land Cover Mapping. In: Proceedings of the 18th Italian Workshop on Neural Networks, WIRN 2008. IOS Press, Amsterdam (2008)
18. Turk, M., Pentland, A.: Eigenfaces for recognition. Journal of Cognitive Neuroscience 3, 71–86 (1991)

Impact of Video Source Coding on the Performance of Stereo Matching

Waqar Zia, Michel Sarkis, and Klaus Diepold

Abstract. In several applications like Telepresence, multi-view video has to be source coded and transmitted to a location where stereo matching is performed. The effect of coding strategies on stereo matching techniques is, however, not thoroughly studied. In this paper, we present an analysis that demonstrates the impact of video source coding on the quality of stereo matching. We use MPEG-4 advanced simple profile to compress the transmitted video and examine the quantitative effect of lossy compression on several types of stereo algorithms. The results of the study are able to show us the relation between video coding compression parameters and the quality of the obtained disparity maps. These results can guide us to select suitable operating regions for both stereo matching and video compression. Thus, we are able to provide a better insight on how to trade between the compression parameters and the quality of dense matching to obtain an optimal performance.

1 Introduction

Telepresence and teleaction (TPTA) is the experience of being present and active at a location distant from one's physical location. In this scenario, a *teleoperator* is a mobile robot equipped with a stereo camera and is placed at a remote location, see Figure 1.

The human operator, situated at a different location, interacts in the remote scene of the teleoperator by means of a generated virtual 3D environment. Audio, video, haptic, and control communications between the sites enable the virtual immersion of the human operator in the teleoperator site. The bandwidth limitation of communication channel makes source-coding of all transmitted data necessary. In addition, constructing the realistic 3D views depicted by the teleoperator involves stereo

Waqar Zia · Michel Sarkis · Klaus Diepold
Institute for Data Processing, Technische Universität München
e-mail: {waqar.zia,michel,kldi}@tum.de

H. Ritter et al. (Eds.): Human Centered Robot Systems, COSMOS 6, pp. 131–140.
springerlink.com

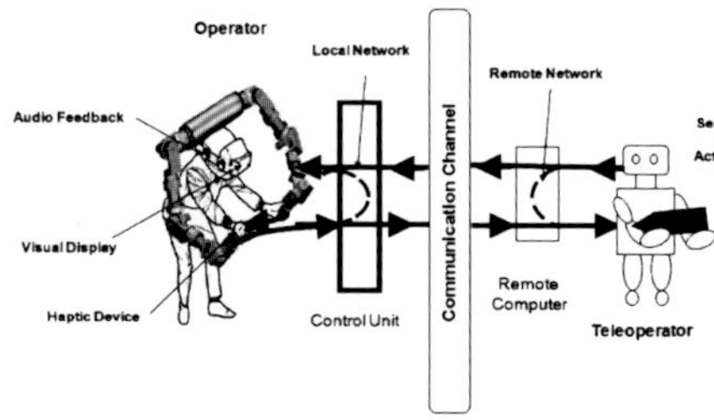

Fig. 1 A typical TPTA scenario

matching of the multi-view images. Stereo matching requires a significant amount of computational resources. In TPTA and other similar applications like autonomous vehicles [4], the processing capacity is quite limited at the site of video capturing. Therefore, it is necessary that stereo reconstruction is performed at the human operator site where more resources are available.

To perform stereo matching, it is necessary to first compute some costs among the possible matches between the images. Then, an energy function is optimized based on the computed costs. A good survey about this topic is found in [13] while various published work are available on the Middlebury webpage in [1]. Algorithms based on dynamic programming (DP) [5, 16] and belief propagation (BP) [7, 17, 10] are among the most well-known techniques implemented since they can be formulated for applications where real-time performance is needed [5, 16, 17].

Stereo matching methods usually assume that the quality of the input images is high. This assumption is unfortunately not valid in a TPTA scenario since matching has to be performed after coding/decoding of the transmitted video content. Interestingly, video coding standards like MPEG-4 advanced simple profile (ASP) or H.264/AVC are designed for lossy compression based on human psycho-visual (HPV) model. The impact of such techniques on the performance of computer vision algorithms in general and specifically stereo matching is not yet thoroughly studied.

Much of the work relevant to the transmission of multi-view images in conjunction with stereo-matching has been done in the area of 3D TV. The 3D TV applications have much different constraints than TPTA. In contrast to TPTA, the resources at the transmission end are higher compared to the receiving end. Therefore, the research focus in 3D TV deals with stereo-matching at the transmitting end, e.g. [6, 9].

In this work, we present a study on the performance of stereo matching techniques when the video has been subjected to lossy compression. We model and simulate the end-to-end system of the combination of video source coding and stereo matching in TPTA. We evaluate several stereo matching algorithms using diverse source content under varying compression parameters. We analyze our results to highlight the impact of source coding of video on stereo matching. The rest of this paper is divided as follows. We discuss the system architecture in Section 2 with a detailed description of its individual components. We present the simulation results in Section 3 followed by some concluding remarks in Section 4.

2 Evaluation Framework

To evaluate the performance of a stereo matching scheme in TPTA, we developed the simulation setup shown in Figure 2. The individual components of this test bench along with their interactions are explained in the sequel. We will conduct the study on stereo videos since the results can be easily extended for more than two views.

2.1 Framework Components

Test sequences: One of the most suitable test content for stereo matching are the Middlebury test images [13, 14]. Although these are still images, we can employ intra-only coding for this content. Such coding is also suitable for minimal complexity as no motion compensation and reconstruction is required at the teleoperator site.

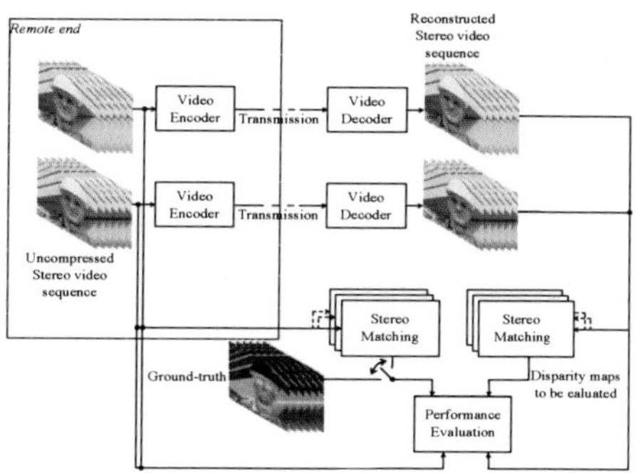

Fig. 2 Block diagram of the simulation framework

Many other sequences associated to multi-view H.264/AVC video coding [8] are available. However, their content is usually not suited for stereo matching because they have been captured in medium to dim indoor lighting conditions and have a significant motion-blur. One commonly used video sequence that we will employ in our tests is "Vassar" [15]. It has been captured in ambient day light and contain no discernable motion blur on the boundaries of the moving objects. Another set we will use is the "Hall" stereo sequence which we captured ourselves [2].

Video encoder: Raw video data needs a large bandwidth when transmitted over a communication channel. The available video compression algorithms like MPEG-4 ASP or H.264/AVC uses the HPV model to compress the data and are not designed for computer vision algorithms. Hence, the distortions introduced by this lossy compression will certainly have an impact on stereo matching. This is why it is necessary and important to visualize the relation between the compression rate of video coding and the quality of stereo matching.

In our tests, we select the MPEG-4 ASP since it holds a unique feature of out-of-loop post-processing (deblocking) filter as opposed to the in-loop deblocking filter in H.264/AVC. This feature reduces the processing overhead of the encoder significantly [12]. We have used the open source codec implementation in our simulations, available in Xvid [3].

Stereo matching: We use in our tests a variety of stereo-matching techniques. We have used the method of [5] which is based on DP. Its simple matching costs that make its application in real-time scenarios like TPTA possible. The second method is the real-time DP algorithm of [16] that employs the high quality adaptive weighting approach described in [18] to compute the matching costs. Both of these techniques are scanline based. The third method we will test is derived in [7] and is based on the belief propagation (BP) optimization principle. We choose to test BP since a lot of research has been done lately to make it applicable in real-time scenarios [17]. As a reference we have also including the Graph Cuts (GC) scheme [11]. Both BP and GC are global techniques.

Performance evaluation: Quantitative evaluation is required for both the received video sequences and the estimated disparity maps. To measure the quality of the received video with respect to the uncompressed version, we use the peak signal to noise ratio (PSNR). To evaluate the quality of the estimated disparity maps from the compressed images, we use the offline evaluation software of [1]. We measure the quality using the percentage of bad pixels (PBP) of the reconstructed disparity map. It is defined for a $M \times N$ image as

$$\mathrm{PBP} = \frac{\sum_{x=1}^{M} \sum_{y=1}^{N} f(d_{x,y}, \hat{d}_{x,y})}{M \times N} \%, f(d_{x,y}, \hat{d}_{x,y}) = 1 \text{ if } |d_{x,y} - \hat{d}_{x,y}| > C, \text{ else } 0 \quad (1)$$

where $\hat{d}_{x,y}$ is the disparity at a pixel location in the reconstructed disparity map from the compressed images and $d_{x,y}$ is the corresponding disparity in the ground truth. C is the pre-defined threshold. The PBP is evaluated for the non-occluded (NON OCCL) regions in the image, to the discontinuity regions (D DISCNT) and to all the regions (ALL). Note that for the other sequences we used, i.e. “Vassar” and “Hall”, there are no ground truth disparity maps available. To compute the PBP in this case, we use the disparity map estimated from the uncompressed video sequence as a reference instead.

2.2 Simulation Environment

The block diagram of the simulation environment is shown in Figure 2. At the encoder, the stereo sequences are rectified using the camera parameters. The MPEG-4 ASP only accepts YUV 4:2:0 at input while the image size must be a multiple of 16 pixels both in height and width. Therefore, the color conversions are done before coding and the dimensions are matched by padding the image with grey values to avoid generating redundant data. The resulting sequences are then fed to a pair of video encoders. Each of the left and right stereo sequences is allowed half of the video bandwidth.

At the receiver site, the compressed video streams are decoded to get the reconstructed video that will be used in stereo matching. Following this, the performance evaluation is done. The PSNR of the reconstructed video is measured. There can be two options to evaluate the disparity-maps, determined by whether a known reference disparity-map ("ground truth") is available or not. If yes, the ground truth is used for performance evaluation. In this case, the performance metrics will consist of the cumulative effect of the coding losses and the algorithm's own performance impairment. If ground truth is not available then the disparity-map generated by the same algorithm is used as a reference, assuming ideal transmission (no video coding losses). In this case the performance metrics will give the exclusive effect of video coding losses. In order to make a fair comparison between the two cases, for the first case, loss-less performance $\mathrm{PBP_r}$ (quality of the disparity map generated by the algorithm without any video compression distortion) will also be indicated along the readings. This curve helps visualizing algorithm's own performance impairment in the overall performance impairment curve.

The Lagrangian-based rate-distortion minimization problem for video coding is as follows. For all the blocks b in a given access unit (e.g. a frame in video sequence), the rate (r) distortion (d) minimization problem is defined as

$$\forall_b \quad m_b^* = \arg\min_{m \in \mathscr{O}} \left(d_{b,m} + \lambda_{\mathscr{O}} r_{b,m} \right). \tag{2}$$

The minimization is done for the usable option set $\mathscr{O}$ with a Lagrangian multiplier $\lambda_{\mathscr{O}}$. However, buffering is possible in most applications that allows variations in $r_{b,m}$ for a given access unit. Hence, multiple solutions of $\lambda_{\mathscr{O}}$ exist that satisfy the constrained minimization. The main target of this study is to understand the relation between the cost function of RD minimization (d) and that of stereo matching algorithms. Since both minimizations have different targets and constraints, it is evident that the two are not linearly related. A single solution to RD minimization for an access unit is not a sufficient measure to highlight the dependence between the two cost functions. In order to achieve higher statistical significance, several solutions are achieved by allowing a maximum buffering of 10 ms. The minimization is started at different blocks within an access unit and is applied to all blocks in a loop-around fashion. Each different start gives a unique solution. The readings are then averaged for at least 128 such variants in order to have a high statistical significance. In case of still images like the Middlebury data set, an image is continuously repeated to yield a "video sequence". The usable option set $\mathscr{O}$ in this case consists only intra-coding options since the predictive coding will give no residual data (nothing to transmit). Intra-only coding for this data set is still relevant for the target application: no motion estimation/compensation is performed at the transmitting end and hence this configuration represents the lowest complexity at the teleoperator site.

3 Results and Discussion

The framework discussed in the previous section is used for detailed performance evaluation. From a system design point of view, the error plot versus the bitrate used to compress a video sequence is most meaningful since the bandwidth is the actual concern. Since the available video sequences have different dimensions, it is more meaningful to normalize the transmission bitrate with the pixel transmission rate. We will designate in the results the algorithm of [5] by SM , the technique of [7] by BP, the method of [16] by CDP and the scheme of [11] by GC. The parameter C for calculating percentage of bad pixels was set to 0, since the calculated disparity maps have integral values and a difference of even one level of disparity might translate in a large and unacceptable variation of the calculated depth.

For each of the still image set (Middlebury), The percentage of bad pixels (PBP) will be plotted against the increasing transmission rate. Moreover, PBP_r is shown at the right of each graph, marked as "loss-less". It should be noted that the coding of uncompressed video content requires 24 bits/pixel.

The results on "Venus" are shown in Figure 3. It can be seen that below a transmission rate of 0.75 bit/pixel, all the algorithms have a very bad performance. Above 1 bit/pixel, the techniques converge to their best performance. For larger bitrate, GC shows the best performance, followed by BP, CDP and SM respectively. However, at low bitrates e.g. at 0.5 bits/pixel, SM is performing similar to BP. Also, unlike CDP and BP, above 1.5 bits/pixel SM shows no further performance improvement.

Instead of bits/pixel, plotting can be done versus PSNR as shown in Figure 5(a). This leads to a more direct relation between the performance of stereo matching and video quality. An interesting observation in this figure is the non-linearity in the vicinity of 0.5 bit/pixel or approximately 33 dB. This arises because of the non-linearities in the matching costs used in SM which are based on the interpolation of pixel discontinuities, see [5]. To emphasize more on that, Figure 5(b) shows the percentage of false edge detections with increasing bitrate. Ironically, below 0.75 bit/pixel, this percentage starts decreasing. This is due to deblocking filter of MPEG-4 ASP which kicks-in at low bitrates to smooth out blocking artifacts, and it is in fact the blocking artifacts that trigger false discontinuities in the matching costs of SM. This can be explained since the video codec focuses on the HPV model based on subjective evaluation. At higher bitrates the filter can cause unpleasant blurring

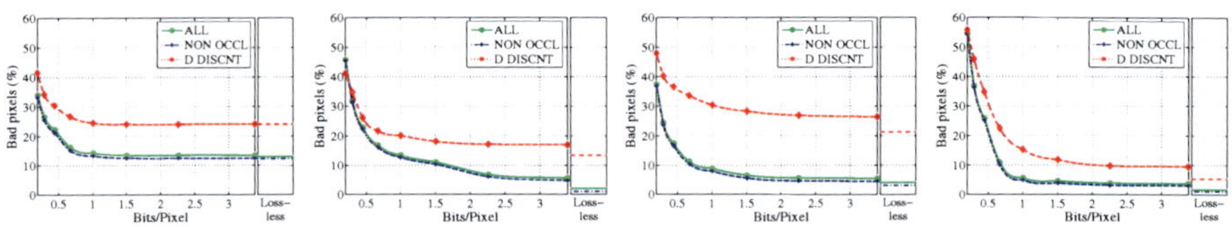

(a) Performance of SM. (b) Performance of BP. (c) CDP. (d) Performance of GC.

Fig. 3 Results for "Venus", in the left part of each plot PBP is plotted versus the transmission rate. The right (smaller) one is the PBP_r.

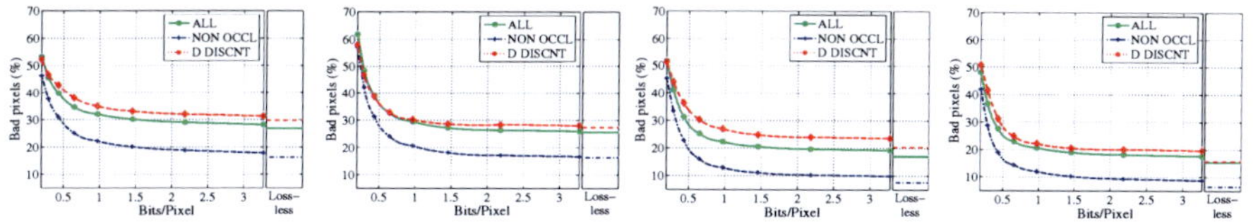

(a) Performance of SM. (b) Performance of BP. (c) CDP. (d) Performance of GC.

Fig. 4 Results for "Cones", in the left part of each plot PBP is plotted versus the transmission rate. The right (smaller) one is the PBP_r.

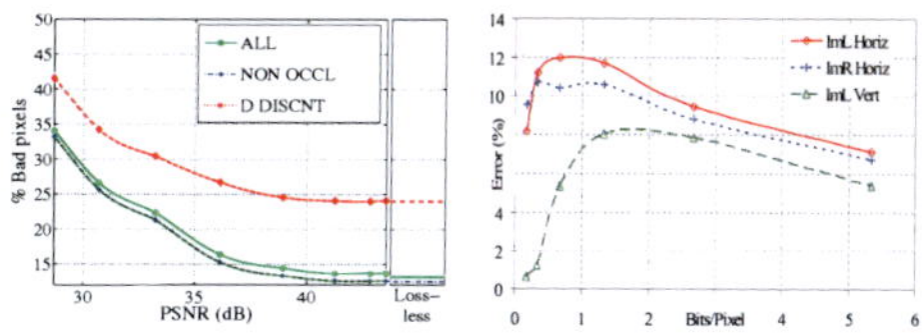

(a) (b)

Fig. 5 (a): Performance of SM on "Venus". (b): Behavior of internal edge detection parameters of SM versus the video transmission rate.

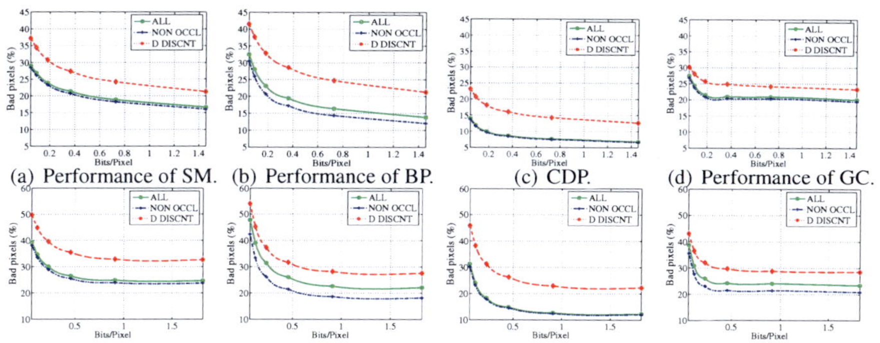

(a) Performance of SM. (b) Performance of BP. (c) CDP. (d) Performance of GC.

(e) Performance of SM. (f) Performance of BP. (g) CDP. (h) Performance of GC.

Fig. 6 Performance results on "Vassar" in the upper row and "Hall" in the lower row

effect. Keeping in view this HPV consideration, MPEG-4 ASP stops using it for higher bitrates.

Figure 4 shows the performance of the stereo algorithms on "Cones". One can notice that the content of this image is much more challenging for stereo matching algorithms as compared to "Venus". The schemes continue to gain some performance even above 2 bits/pixel. Still GC seems to perform the best, followed by CDP, BP and SM respectively. It is interesting to observe that BP performs worse than CDP even with its parameters set to the values applied in [7]. The results of

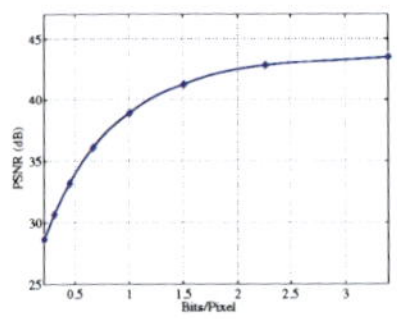

(a) "Venus"

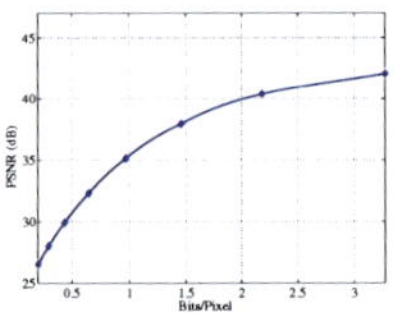

(b) "Cones"

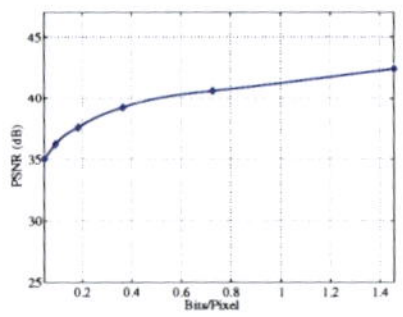

(c) "Vassar"

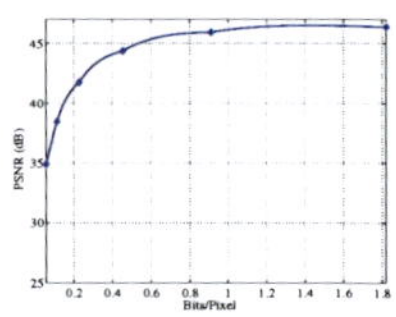

(d) "Hall"

Fig. 7 Rate-Distortion curves for the sequences used

(a) Teddy original.

(b) CDP disparity-map

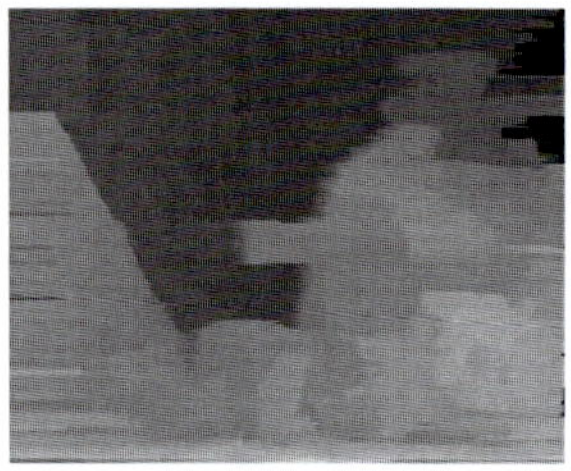

(c) CDP disparity-map after compression

(d) Cones original.

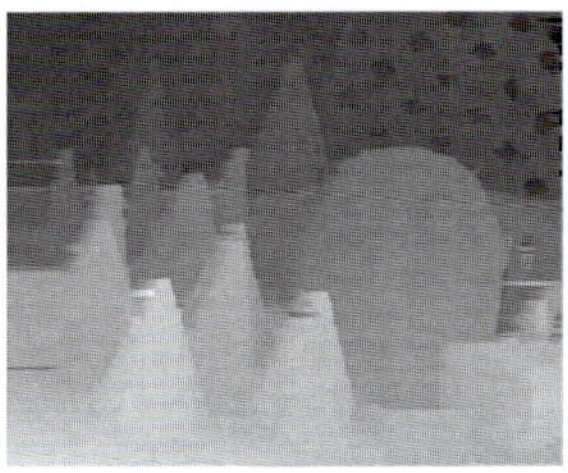

(e) CDP disparity-map

(f) CDP disparity-map after compression

Fig. 8 Sample images and disparity maps using CDP

"Cones" were not presented in the aforementioned work. Below 0.5 bits/pixel, both SM and BP behave in a similar fashion.

The results of the last two video sequences we will be using are shown in Figure 6. It can be seen in "Vassar" (first row of the figure) that below 0.6 bits/pixel, CDP shows much superior performance to BP. At the highest bitrate shown here, the algorithms still seem to gain in performance. It is worth to note that even with well-tuned parameters, GC performs worst in these sequences. The "Hall" shows similar results, except that there seems no significant improvement in performance above 1 bit/pixel. Here as well, GC performs relatively in a poor manner. By looking at the RD curves in Figure 7, one can see that the curve of "Hall" is almost flattened above this rate. The difference in the shape of RD curves of "Vassar" and "Hall" is due to the difference in scene content as well as the RD-minimization algorithm of the video codec. Also, the content has higher noise compared to the Middlebury images which are generated in a carefully controlled environment.

For a visual comparison, Figure 8 shows a reference image of "Cones" and "Teddy" along with the disparity maps achieved by CDP in a lossless scenario. On the very right is the disparity map obtained with CDP after compression at 33 dB. This shows some visual artifacts especially in the non-textured regions and around edges.

4 Conclusion and Outlook

In this study, we have modeled and simulated the video source coding and stereo matching in a TPTA scenario. MPEG-4 ASP is used for stereo video compression. Several stereo matching algorithms are evaluated against a variety of video sequences and coding conditions. A wide range of results were obtained that yield a few important observations. It can be first concluded that a compression factor of 24 or less in general should be selected to avoid a large performance degradation. Moreover, CDP shows extremely good performance across the entire transmission rate spectrum due to the used matching costs. However, it requires far more computational effort when compared to SM. Detailed analysis also shows that the matching costs in SM could benefit from the post-processing of MPEG-4 ASP. Unfortunately, this cannot be done at higher bitrates because of the HPV model considerations. Global techniques like BP and GC show extremely good performance for carefully generated low noise content, but noise affects their performance in the most adverse way. As a future work, we will assess the H.264/AVC multi-view codec as a candidate source coder while addressing its complexity issues.

References

1. Middlebury stereo evaluation, `http://vision.middlebury.edu/stereo/data`
2. Sequence "hall", `http://www.ldv.ei.tum.de/Members/waqar/Hall.avi`
3. Xvid open-source research project, `http://www.xvid.org`
4. Achtelik, M.: Vision-Based Pose Estimation for Autonomous Micro Aerial Vehicles. Master's thesis, Technische Universität München, Munich, Germany (2009)
5. Birchfield, S., Tomasi, C.: Depth discontinuities by pixel-to-pixel stereo. Int. J. Computer Vision 35(3), 269–293 (1999)
6. Ekmekcioglu, E., Worrall, S.T., Kondoz, A.M.: Bit-rate adaptive downsampling for the coding of multi-view video with depth information. In: 3DTV-Conf. (2008)
7. Felzenszwalb, P.F., Huttenlocher, D.P.: Efficient belief propagation for early vision. Int. J. Computer Vision 70(1), 41–54 (2006)
8. ISO/IEC JTC1/SC29/WG11 N7327: Call for proposals on multi-view video coding (2005)

9. Karlsson, L.S., Sjöström, M.: Region-of-interest 3d video coding based on depth images. In: 3DTV-Conf. (2008)
10. Klaus, A., Sormann, M., Karner, K.: Segment-based stereo matching using belief propagation and a self-adapting dissimilarity measure. In: Int. Conf. Pattern Recog. (2006)
11. Kolmogorov, V., Zabih, R., Gortler, S.: Multi-camera scene reconstruction via graph cuts. In: Heyden, A., Sparr, G., Nielsen, M., Johansen, P. (eds.) ECCV 2002. LNCS, vol. 2352, pp. 82–96. Springer, Heidelberg (2002)
12. Saponara, S., Blanch, C., Denolf, K., Bormans, J.: The jvt advanced video coding standard: Complexity and performance analysis. In: Packet Video Workshop (2003)
13. Scharstein, D., Szeliski, R.: A taxonomy and evaluation of dense two-frame stereo correspondence algorithms. Int. J. Computer Vision 47(1), 7–42 (2002)
14. Scharstein, D., Szeliski, R.: High-accuracy stereo depth maps using structured light. In: IEEE Conf. Computer Vision and Pattern Recognition, pp. 195–202 (2003)
15. Vetro, A., McGuire, M., Matusik, W., Behrens, A., Lee, J., Pfister, H.: Multiview video test sequences from merl. ISO/IEC JTC1/SC29/WG11 Document m12077 (2005)
16. Wang, L., Liao, M., Gong, M., Yang, R., Nistér, D.: High quality real-time stereo using adaptive cost aggregation and dynamic programming. In: Int. Symp. 3D Data Processing, Visualization and Transmission, pp. 798–805 (2006)
17. Yang, Q., Wang, L., Yang, R., Wang, S., Liao, M., Nistér, D.: Real-time global stereo matching using hierarchical belief propagation. In: Br. Machine Vision Conf., pp. 989–998 (2006)
18. Yoon, K.J., Kweon, I.-S.: Adaptive support-weight approach for correspondence search. IEEE Trans. Pattern Analysis and Machine Intelligence 28(4), 650–656 (2006)

3D Action Recognition in an Industrial Environment

Markus Hahn, Lars Krüger, Christian Wöhler, and Franz Kummert

Abstract. In this study we introduce a method for 3D trajectory based recognition of and discrimination between different working actions. The 3D pose of the human hand-forearm limb is tracked over time with a two-hypothesis tracking framework based on the Shape Flow algorithm. A sequence of working actions is recognised with a particle filter based non-stationary Hidden Markov Model framework, relying on the spatial context and a classification of the observed 3D trajectories using the Levenshtein Distance on Trajectories as a measure for the similarity between the observed trajectories and a set of reference trajectories. An experimental evaluation is performed on 20 real-world test sequences acquired from different viewpoints in an industrial working environment. The action-specific recognition rates of our system correspond to more than 90%. The actions are recognised with a delay of typically some tenths of a second. Our system is able to detect disturbances, i.e. interruptions of the sequence of working actions, by entering a safety mode, and it returns to the regular mode as soon as the working actions continue.

1 Introduction

Today, industrial production processes in car manufacturing worldwide are characterised by either fully automated production sequences carried out solely by industrial robots or fully manual assembly steps where only humans work together on the same task. Up to now, close collaboration between humans and machines, especially industrial robots, is very limited and usually not possible due to safety

Markus Hahn · Lars Krüger · Christian Wöhler
Daimler AG, Group Research and Advanced Engineering, P.O. Box 2360, D-89013 Ulm, Germany
e-mail: {Markus.Hahn,Lars.Krueger,Christian.Woehler}@daimler.com

Christian Wöhler · Franz Kummert
Applied Informatics, Faculty of Technology, Bielefeld University, Universitätsstraße 25, D–33615 Bielefeld, Germany
e-mail: franz@techfak.uni-bielefeld.de

H. Ritter et al. (Eds.): Human Centered Robot Systems, COSMOS 6, pp. 141–150.
springerlink.com

concerns. Industrial production processes can increase efficiency by establishing a close collaboration of humans and machines exploiting their unique capabilities. A safe interaction between humans and industrial robots requires vision methods for 3D pose estimation, tracking, and recognition of the motion of both human body parts and robot parts.

Previous work in the field of human motion capture and recognition is extensive. Moeslund et al. [12] give a detailed introduction and overview. Bobick and Davis [2] provide another good introduction. They classify human motion using a temporal template representation from a set of consecutive background-subtracted images. A drawback of this approach is its dependence on the viewpoint.

Li et al. [11] use Hidden Markov Models (HMMs) to classify hand trajectories of manipulative actions and take into account the object context. In [3] the motion of head and hand features is used to recognise Tai Chi gestures by HMMs. Head and hand are tracked with a real-time stereo blob tracking algorithm. HMMs are used in many other gesture recognition systems due to their ability to probabilistically represent the variations of the training data. Dynamic Bayesian Networks (DBN) generalise HMMs and are able to consider several random variables [13]. In [14] two-person interactions are recognised with a DBN. The system has three abstraction levels. On the first level, human body parts are detected using a Bayesian network. On the second level, DBNs are used to model the actions of a single person. On the highest level, the results from the second level are used to identify the interactions between individuals.

A well known approach to gesture recognition is the method by Black and Jepson [1], who present an extension of the CONDENSATION algorithm and model gestures as temporal trajectories of the velocity of the tracked hands. They perform a fixed size linear template matching weighted by the observation densities. Fritsch et al. [5] extend their work by incorporation of situational and spatial context. Both approaches merely rely on 2D data. Hofemann [8] extends the work in [5] to 3D data by using a 3D body tracking system. The features used for recognition are the radial and vertical velocities of the hand with respect to the torso.

Croitoru et al. [4] present a non-iterative 3D trajectory matching framework which is invariant to translation, rotation, and scale. They introduce a pose normalisation approach which is based on physical principles, incorporating spatial and temporal aspects of trajectory data. They apply their system to 3D trajectories for which the beginning and the end is known. This is a drawback for applications processing a continuous data stream, since the beginning and end of an action are often not known in advance.

This paper addresses the problem of tracking and recognising the motion of human body parts in a working environment, which is a precondition for a close collaboration between human workers and industrial robots. Our 3D tracking and recognition system consists of three main components: the camera system, the model-based 3D tracking system, and the trajectory-based recognition system. As an imaging system we use a calibrated small-baseline trinocular camera sensor similar to that of the SafetyEYE protection system (www.safetyeye.com) which is used in our production processes to protect human workers.

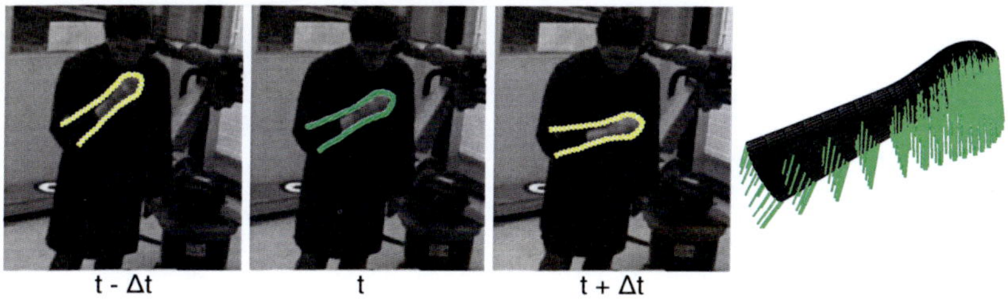

Fig. 1 Example of a spatio-temporal 3D pose estimation result

2 The 3D Tracking System

We rely on the spatio-temporal 3D pose estimation and tracking system introduced in [7], which is based on the Shape Flow algorithm. A spatio-temporal 3D model of the human hand-forearm limb is used as shown in Fig. 1 (right), made up by a kinematic chain connecting the two rigid elements forearm and hand. The model consists of five truncated cones and one complete cone. The Shape Flow algorithm fits the parametric curve to multiple calibrated images by separating the grey value statistics on both sides of the projected curve. Fig. 1 illustrates a correspondingly obtained spatio-temporal 3D pose estimation result. At time step t the projected contour of the estimated 3D pose is shown as a green curve, while the images at time steps $t \pm \Delta t$ depict the projected contours inferred from the temporal pose derivative as yellow curves. To start tracking, a coarse initialisation of the model parameters at the first time step is required. In the tracking system we apply two instances of the Shape Flow algorithm. A winner-takes-all component selects the best-fitting spatio-temporal model at each time step using different criteria. A more detailed description is given in [7].

3 Recognition System

The working action recognition system is based on a 3D trajectory classification and matching approach. The tracking stage yields a continuous data stream of the 3D pose of the tracked hand-forearm limb. Our trajectories are given by the 3D motion of the wrist point. The cyclic sequence of working actions in an engine assembly scenario is known to our system. However, it may be interrupted by "unknown" motion patterns. The beginning and the end of a trajectory are not known a priori, which is different from the setting regarded in [4]. To allow an online action recognition in the continuous data stream, we apply a sliding window approach which enables the system to perform a recognition of and discrimination between human motion patterns.

Due to the fact that our system is designed for safe human–robot interaction, we implemented a recognition stage with two levels (Fig. 2). At the first level, the decision is made whether the human worker performs a known working action (regular

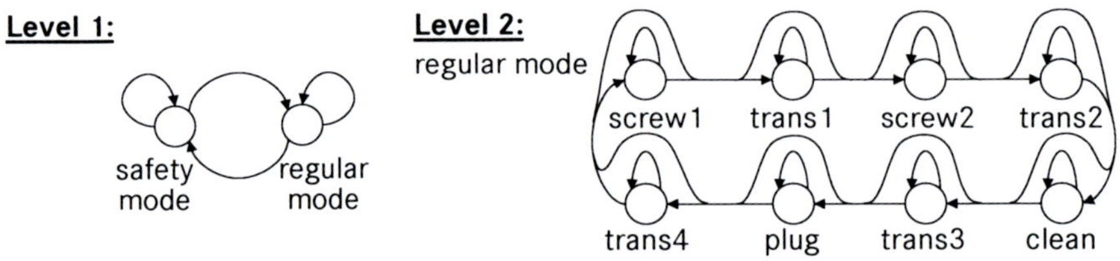

Fig. 2 The two-level architecture of the safety system

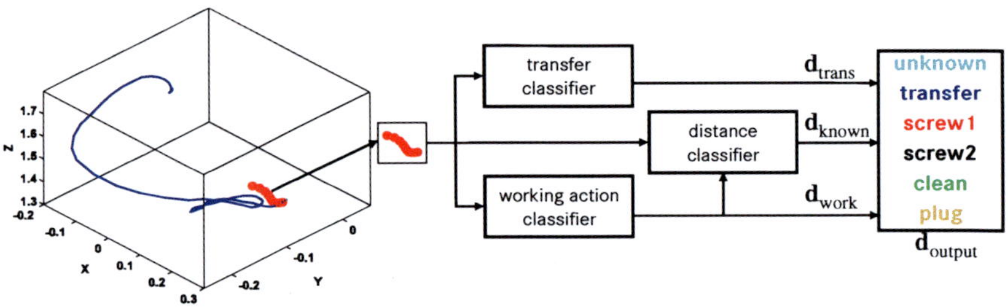

Fig. 3 Classifiers and smoothed 3D input trajectory. The values in the current sliding window are marked in red, all others in blue.

mode) or an unknown motion (safety mode) based on a set of trajectory classifiers (cf. Section 3.1). In the safety mode (level 1), the system may prepare to slow down or halt the industrial robot. The regular mode (level 2) defines the cyclic working process performed by the human worker. It is implemented as a HMM in which the state is continuously estimated by a particle filter, where the particle weights are computed with the trajectory classifiers described in Section 3.1 and a trajectory matching approach based on the Levenshtein Distance on Trajectories (LDT) measure [6].

3.1 Trajectory Classifiers

The state of system level 1 according to Fig. 2 is determined by a set of classifiers based on features extracted from the trajectory data in the sliding window. In a preprocessing step, noise and outliers in the trajectory data are reduced by applying a Kalman Filter based smoothing procedure [4]. The kinematic model of the Kalman Filter is a constant-velocity model. Fig. 3 (left) depicts the smoothed 3D trajectory (blue) and the 3D points (red) in the current sliding window of size $W = 8$ time steps. We apply two second-order polynomial classifiers and a Mahalanobis distance classifier [15] in a hierarchical manner (cf. Fig. 3). We found experimentally that a polynomial classifier is favourable for classifying transfer motion patterns and working actions based on small training sets (cf. Section 4), while distance-based classifiers turned out to be too restrictive for these tasks when applied to our large test set. The Mahalanobis distance classifier was used for post-processing the outputs of the polynomial classifiers. At time step t the current input trajectory $\mathbf{Z}_t$ with

$$\mathbf{Z}_t = \left[(X_{(t-W+1)}, Y_{(t-W+1)}, Z_{(t-W+1)})^T, \; \ldots, (X_t, Y_t, Z_t)^T \right], \tag{1}$$

consisting of the last W elements of the continuous data stream, is used to extract features to which the set of classifiers is applied as described in the following.

Motion patterns occurring between two working actions (here denoted by "transfer motion") are recognised by a polynomial classifier using two features, (i) the travelled distance along the trajectory and (ii) the maximum angle between two consecutive motion direction vectors in the sliding window. The output discriminant vector $\mathbf{d}_{\text{trans}}$ of this classifier consists of the two classes "transfer motion" and "no transfer motion". Normalisation yields the transfer discriminant vector $\tilde{\mathbf{d}}_{\text{trans}}$ of unit length with components in the interval $[0,1]$.

Since it is known where the worker has to tighten a screw or to fit a plug, the second polynomial classifier is used for recognising working actions by incorporating spatial context for the actions "tighten screw 1", "tighten screw 2", "clean" and "plug". These 3D positions are constant across the sequence and are obtained based on the known 3D pose of the engine. As features, the polynomial classifier for working actions uses the minimum distance in the sliding window to the 3D position of (i) screw 1, (ii) screw 2, (iii) the area to be cleaned, and (iv) the position where to fit the plug. Normalisation yields the working action discriminant vector $\tilde{\mathbf{d}}_{\text{work}}$.

The Mahalanobis distance classifier is applied to the result of the polynomial classifier for working actions and decides whether the recognised working action is a known one, since such motion patterns can only occur close to the 3D object associated with that action. The classifier applies a winner-takes-all approach to the discriminant vector $\tilde{\mathbf{d}}_{\text{work}}$ and performs a comparison between the training data and the measured distance to the 3D object associated with the winner class. The result is the normalised discriminant vector $\tilde{\mathbf{d}}_{\text{known}}$.

Based on the normalised discriminant vectors of the three classifiers, which are given by

$$\tilde{\mathbf{d}}_{\text{trans}} = \begin{pmatrix} \tilde{d}_{\text{trans}} \\ 1 - \tilde{d}_{\text{trans}} \end{pmatrix}, \quad \tilde{\mathbf{d}}_{\text{known}} = \begin{pmatrix} \tilde{d}_{\text{known}} \\ 1 - \tilde{d}_{\text{known}} \end{pmatrix}, \quad \tilde{\mathbf{d}}_{\text{work}} = \begin{pmatrix} \tilde{d}_{\text{screw1}} \\ \tilde{d}_{\text{screw2}} \\ \tilde{d}_{\text{clean}} \\ \tilde{d}_{\text{plug}} \end{pmatrix}, \tag{2}$$

an overall discriminant vector $\mathbf{d}_{\text{output}}$ for the six classes "unknown", "transfer", "screw 1", "screw 2", "clean", and "plug" is determined, which is given by

$$\mathbf{d}_{\text{output}} = \begin{pmatrix} d_{\text{unknown}} \\ d_{\text{trans}} \\ d_{\text{screw1}} \\ d_{\text{screw2}} \\ d_{\text{clean}} \\ d_{\text{plug}} \end{pmatrix} = \begin{pmatrix} 1 - \tilde{d}_{\text{known}} \\ \tilde{d}_{\text{known}} \cdot \tilde{d}_{\text{trans}} \\ \tilde{d}_{\text{known}} \cdot (1 - \tilde{d}_{\text{trans}}) \cdot \tilde{d}_{\text{screw1}} \\ \tilde{d}_{\text{known}} \cdot (1 - \tilde{d}_{\text{trans}}) \cdot \tilde{d}_{\text{screw2}} \\ \tilde{d}_{\text{known}} \cdot (1 - \tilde{d}_{\text{trans}}) \cdot \tilde{d}_{\text{clean}} \\ \tilde{d}_{\text{known}} \cdot (1 - \tilde{d}_{\text{trans}}) \cdot \tilde{d}_{\text{plug}} \end{pmatrix}. \tag{3}$$

Finally, normalisation of $\mathbf{d}_{\text{output}}$ to unit length yields the classification vector $\tilde{\mathbf{d}}_{\text{output}}$.

3.2 Recognition of the Sequence of Working Actions

The decision whether the system is in safety mode or in regular mode is made based on the first discriminant value $\tilde{d}_{\mathrm{unknown}}$ of the normalised output discriminant vector $\tilde{\mathbf{d}}_{\mathrm{output}}$ and the matching accuracy of the particle weights in system level 2, where the observed trajectories are analysed with respect to the occurrence of known working actions.

Similar to [11] we apply a particle filter based non-stationary HMM matching in order to recognise the sequence of working actions. The HMM of system level 2 (Fig. 2) is derived from the known cyclic working task, defined by a parameter set $\lambda = (S, A, B, \Pi)$:

- $S = \{q_1, \ldots, q_n\}$, the set of hidden states;
- $A = \{a_{ij,t} | a_{ij,t} = P(q_t = s_j | q_{t-1} = s_i)\}$, non-stationary (time-dependent) transition probability from state s_i to s_j;
- $B = \{b_{i,k} | b_{i,k} = P(o_t = v_k | q_t = s_i)\}$, probability of observing the visible state v_k given the hidden state s_i;
- $\Pi = \{\pi_i | \pi_i = P(q_1 = s_i)\}$, initial probability of state s_i.

We assigned a set of reference trajectories to each hidden state $\{q_1, \ldots, q_n\}$ based on the associated working action. Our system relies on a small number of reference trajectories which are defined by manually labelled training sequences. To cope with different working speeds, the defined reference trajectories are scaled in the temporal domain (from -20% to $+20\%$ of the total trajectory length).

Similar to [11] the CONDENSATION algorithm [9] is used to estimate the state of the HMM based on temporal propagation of a set of N weighted particles:

$$\left\{(\mathbf{s}_t^{(1)}, w_t^{(1)}), \ldots, (\mathbf{s}_t^{(N)}, w_t^{(N)})\right\} \quad \text{with} \quad \mathbf{s}_t^{(i)} = \left\{q_t^{(i)}, \phi_t^{(i)}\right\}. \tag{4}$$

The particle $\mathbf{s}_t^{(i)}$ contains the hidden state $q_t^{(i)}$ and the current phase $\phi_t^{(i)}$ in this hidden state, where the phase indicates the fraction by which the working action has been completed. The resampling step reallocates a certain fraction of the particles with regard to the predefined initial distribution Π. The weight $w_t^{(i)}$ of a particle is calculated according to $w_t^{(i)} = p(o_t | \mathbf{s}_t^{(i)}) / \sum_{j=1}^{N} p(o_t | \mathbf{s}_t^{(i)})$, where $p(o_t | \mathbf{s}_t^{(i)})$ is the observation probability o_t given the hidden state $q_t^{(i)}$ and its phase $\phi_t^{(i)}$. The propagation of the weighted particles over time consists of a prediction, selection, and update step.

Select: Selection of $N - M$ particles $\mathbf{s}_{t-1}^{(i)}$ according to their respective weight $w_{t-1}^{(i)}$ and random distribution of M new particles over all other states in the HMM.

Predict: The current state of each particle $\mathbf{s}_t^{(i)}$ is predicted based on the selected particles, the HMM structure (Fig. 2), and the current phase $\phi_t^{(i)}$. The transition probabilities A are not stationary but depend on the current phase $\phi_t^{(i)}$ of the particle. The phase is always restricted to the interval $[0, 1]$. A high phase value indicates that the reference trajectories are almost traversed and that there is an increased probability to proceed to the next state.

Update: In the update step, the weights of the predicted particles are computed based on the discriminant vector $\tilde{\mathbf{d}}_{\text{output}}$ derived from the classifiers and a trajectory matching based on the LDT measure [6]. To compute the weight $w_t^{(i)}$ of a particle $\mathbf{s}_t^{(i)}$, the 3D data $\mathbf{Z}_t$ in the current sliding window are matched with the current sub-trajectory of all reference trajectories of the hidden state $q_t^{(i)}$. The current sub-trajectory in a hypothesis trajectory is defined by its phase $\phi_t^{(i)}$ and length W. The weight is given by the LDT measure of the best matching reference trajectory multiplied by the discriminant value in $\tilde{\mathbf{d}}_{\text{output}}$ associated with the corresponding action class of the hidden state $q_t^{(i)}$. It is possible that the same working action is performed at different positions in 3D space, e.g. tightening a screw at different locations of the engine. Hence, the trajectories are normalised w.r.t. translation, rotation, and scaling. For this purpose we apply the quaternion-based approach introduced in [10].

The set of weighted particles yields a likelihood at each time step for being in the specific working action states of the HMM (cf. Fig. 4 (bottom) for an example sequence).

4 Experimental Evaluation

The system is evaluated by analysing 20 trinocular real-world test sequences acquired from different viewpoints. The time interval between subsequent image triples acquired with our small-baseline trinocular camera sensor amounts to $\Delta t =$ 71 ms. These sequences contain working actions performed by eight different test persons in front of a complex cluttered working environment. Each sequence contains at least 300 image triples. The distance of the test persons to the camera system amounts to 2.2–3.3 m. Each sequence contains the working actions listed in Table 1, where all actions are performed with the right hand. All sequences were robustly and accurately tracked at all time steps with the spatio-temporal 3D tracking system described in Section 2.

Only two sequences, each comprising 400 image triples, in which the working actions are performed by two different individuals, are used for training the system. These two individuals (teachers) are well trained while all other test persons are less well trained since they were shown the actions by the teachers only once in advance. This teacher-based approach is motivated by our application scenario, in which workers are generally trained by only a few experts.

We assigned ground truth labels manually to all images of the training and test sequences. All results are obtained with a total number of $N = 500$ particles and $M = 100$ uniformly distributed particles. The computation time of our Matlab implementation of the recognition system is around 1 frame per second on a Core 2 Duo with 2.4GHz.

Table 1 Recognition results on our set of 20 test sequences

	tightening screw1	tightening screw2	cleaning	plugging
Total [#]	26	23	28	32
Correct [#]	24	21	27	30
Duplicate [#]	1	2	9	0
Deletion [#]	2	2	1	2
Substitution [#]	0	0	0	0
Insertion [#]	0	2	1	1
Recognition rate [%]	92.3	91.3	96.4	93.8
Word error rate [%]	7.7	17.4	7.1	9.4
Delay begin, mean [ms]	216	777	1102	364
Delay begin, std [ms]	902	702	1753	666
Delay end, mean [ms]	442	246	−1180	239
Delay end, std [ms]	1076	588	1500	319

Table 1 shows that the system achieves average action recognition rates of more than 90% on the test sequences. The relatively large number of duplicates for the cleaning action is due to the erroneous recognition of short transfer phases during these actions as a result of fast motion. The recognition errors can be ascribed to tracking inaccuracies and to motion patterns that differ in space and time from the trained motion patterns. Higher recognition rates may be achieved by using more training sequences, since the scaling in the temporal domain of the reference trajectories is not necessarily able to cope with the observed variations of the motion patterns. The average word error rate, which is defined as the sum of insertions, deletions, and substitutions, divided by the total number of test patterns, amounts to about 10%. Our recognition rates are similar to those reported by Croitoru et al. [4] and Fritsch et al. [5]. Segmented 3D trajectories are used in [4], which is different from our approach, while the method described in [5] relies on 2D data and is not independent of the viewpoint. A precise and stable 3D tracking is essential for our approach since the 3D positions associated with the working actions are separated from each other by only a few decimetres.

On the average, our system recognises the working actions with a delay of several tenths of a second when compared to the manually labelled ground truth, except for the cleaning action which is typically recognised by about one second earlier (negative mean delay). The standard deviations of the delays are typically comparable to or larger than their mean values. One should keep in mind, however, that our manually assigned labels are not necessarily perfectly accurate.

Beyond the recognition of working actions, our system is able to recognise disturbances, occurring e.g. when the worker interrupts the sequence of working actions by blowing his nose. The system then enters the safety mode (indicated by high values of $\tilde{d}_{\text{unknown}}$ in Fig. 4 (top)) and returns to the regular mode as soon as the working actions are continued.

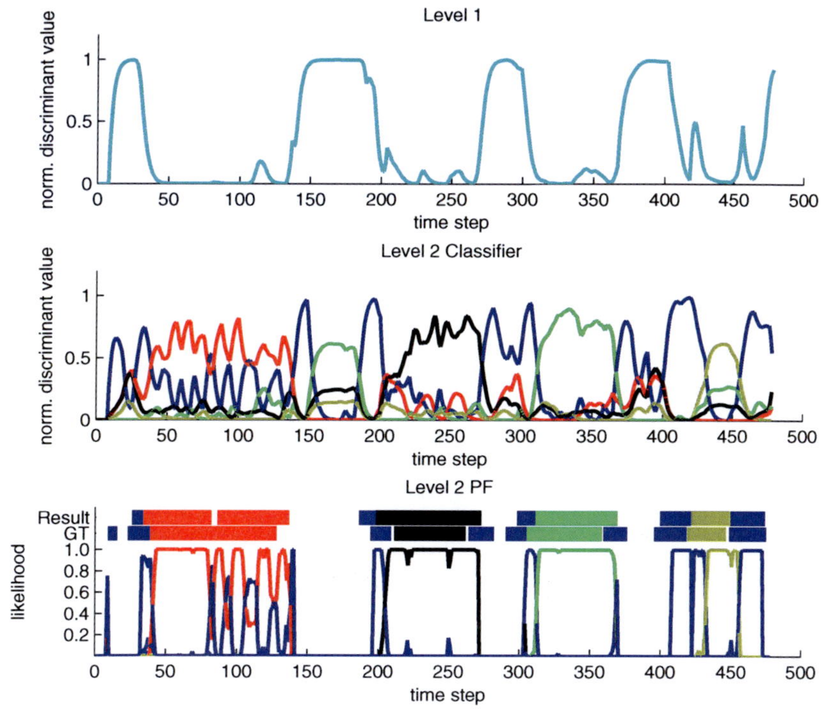

Fig. 4 Recognition of working actions for an example sequence. Top: Normalised classifier output $\tilde{d}_{\mathrm{unknown}}$. Middle: Last five components of the output $\tilde{\mathbf{d}}_{\mathrm{output}}$ (red: screw 1; black: screw 2; green: clean; brown: plug; blue: transfer). Bottom: Final action recognition result compared to ground truth (GT).

5 Summary and Conclusion

In this study we have introduced a method for 3D trajectory based recognition of and discrimination between different working actions. The 3D pose of the human hand-forearm limb has been tracked over time with a two-hypothesis tracking framework based on the Shape Flow algorithm. Sequences of working actions have been recognised with a particle filter based non-stationary HMM framework, relying on the spatial context and a classification of the observed 3D trajectories using the LDT as a measure for the similarity between the observed trajectories and a set of reference trajectories. An experimental evaluation has been performed on 20 real-world test sequences acquired from different viewpoints in an industrial working environment. The action-specific recognition rates of our system correspond to more than 90%, where the actions have been recognised with a delay of typically some tenths of a second. Our system is able to detect disturbances, i.e. interruptions of the sequence of working actions, by entering a safety mode, and it returns to the regular mode as soon as the working actions continue.

An extension of the recognition system to more actions is straightforward. The HMM then needs to be extended by adding new working actions, the classifiers need to be re-trained, and the number of particles should be increased, since the required number of particles scales linearly with the number of actions. Future work may involve online learning of reference trajectories and the definition of a more complex interaction scenario with several human workers and industrial robots. In a such a scenario the usage of more complex DBNs would be appropriate.

References

1. Black, M.J., Jepson, A.D.: A probabilistic framework for matching temporal trajectories: Condensation-based recognition of gestures and expressions. In: Burkhardt, H.-J., Neumann, B. (eds.) ECCV 1998. LNCS, vol. 1406, pp. 909–924. Springer, Heidelberg (1998)
2. Bobick, A.F., Davis, J.W.: The recognition of human movement using temporal templates. IEEE Trans. Pattern Anal. Mach. Intell. 23(3), 257–267 (2001)
3. Campbell, L.W., Becker, D.A., Azarbayejani, A., Bobick, A.F., Pentland, A.: Invariant features for 3-d gesture recognition. In: FG 1996: Proc. of the 2nd Int. Conf. on Automatic Face and Gesture Recognition, FG 1996 (1996)
4. Croitoru, A., Agouris, P., Stefanidis, A.: 3d trajectory matching by pose normalization. In: GIS 2005: Proc. of the 13th annual ACM international workshop on Geographic information systems, pp. 153–162 (2005)
5. Fritsch, J., Hofemann, N., Sagerer, G.: Combining sensory and symbolic data for manipulative gesture recognition. In: Proc. Int. Conf. on Pattern Recognition, vol. 3, pp. 930–933 (2004)
6. Hahn, M., Krüger, L., Wöhler, C.: 3d action recognition and long-term prediction of human motion. In: Gasteratos, A., Vincze, M., Tsotsos, J.K. (eds.) ICVS 2008. LNCS, vol. 5008, pp. 23–32. Springer, Heidelberg (2008)
7. Hahn, M., Krüger, L., Wöhler, C.: Spatio-temporal 3d pose estimation and tracking of human body parts using the shape flow algorithm. In: Proc. Int. Conf. on Pattern Recognition, Tampa, USA (2008)
8. Hofemann, N.: Videobasierte Handlungserkennung für die natürliche Mensch-Maschine-Interaktion. Dissertation, Universität Bielefeld, Technische Fakultät (2007)
9. Isard, M., Blake, A.: Condensation—conditional density propagation forvisual tracking. Int. J. Comput. Vision 29(1), 5–28 (1998)
10. Kearsley, S.K.: On the orthogonal transformation used for structural comparisons. Acta Cryst. A45, 208–210 (1989)
11. Li, Z., Fritsch, J., Wachsmuth, S., Sagerer, G.: An object-oriented approach using a top-down and bottom-up process for manipulative action recognition. In: Franke, K., Müller, K.-R., Nickolay, B., Schäfer, R. (eds.) DAGM 2006. LNCS, vol. 4174, pp. 212–221. Springer, Heidelberg (2006)
12. Moeslund, T.B., Hilton, A., Krüger, V.: A survey of advances in vision-based human motion capture and analysis. Computer Vision and Image Understanding 104(2), 90–126 (2006)
13. Murphy, K.P.: Dynamic bayesian networks: representation, inference and learning. PhD thesis, Chair-Russell, Stuart (2002)
14. Park, S.: A hierarchical bayesian network for event recognition of human actions and interactions. Multimedia Systems, Sp. Iss. on Video Surveillance 10(2), 164–179 (2004)
15. Schürmann, J.: Pattern classification: a unified view of statistical and neural approaches. John Wiley & Sons, Inc., Chichester (1996)

Investigating Human-Human Approach and Hand-Over

Patrizia Basili, Markus Huber, Thomas Brandt, Sandra Hirche, and Stefan Glasauer

Abstract. Humans interact safely, effortlessly, and intuitively with each other. An efficient robot assistant should thus be able to interact in the same way. This requires not only that the robot can react appropriately to human behaviour, but also that robotic behaviour can be understood intuitively by the human partners. The latter can be achieved by the robot mimicking certain aspects of human behaviour so that the human partner can more easily infer the intentions of the robot. Here we investigate a simple interaction scenario, approach and hand-over, to gain better understanding of the behavioural patterns in human-human interactions. In our experiment, one human subject, holding an object, approached another subject with the goal to hand over the object. Head and object positions were measured with a motion tracking device to analyse the behaviour of the approaching human. Interaction indicated by lifting the object in order to prepare for hand-over started approximately 1.2 s before the actual hand-over. Interpersonal distance varied considerably between subjects with an average of 1.16 m. To test whether the behavioural patterns observed depended on two humans being present, we replaced in a second experiment the receiving subject with a table. We found that the behaviour of the transferring subject was very similar in both scenarios. Thus, the presence of the receiving subject plays a minor role in determining parameters such as start of interaction or interaction distance. We aim to implement and test the parameters

Patrizia Basili · Markus Huber · Stefan Glasauer
Center for Sensorimotor Research, Institute of Clinical Neurosciences,
Ludwig-Maximilians-Universität München
e-mail: {p.basili,markus.huber,S.Glasauer}@lrz.uni-muenchen.de

Thomas Brandt
Chair of Clinical Neurosciences, Ludwig-Maximilians-Universität München
e-mail: thomas.brandt@med.uni-muenchen.de

Sandra Hirche
Institute of Automatic Control Engineering, Technische Universität München
e-mail: hirche@lsr.ei.tum.de

H. Ritter et al. (Eds.): Human Centered Robot Systems, COSMOS 6, pp. 151–160.
springerlink.com

derived experimentally in a robotic assistant to improve and facilitate human-robot interaction.

1 Introduction

According to the human-centered approach to robotics, efficient human-robot interaction can be achieved by the robot mimicking crucial aspects of human behaviour [5]. This enables the human partners to intuitively infer the intentions of the robot and thus to predict its further actions. An interaction between humans can be described by a set of algorithms which allow a natural, effective and safe interaction. For a robot, this set needs to be specified. The aim is therefore to define and to implement mathematical models derived from human-human experiments in autonomous robots. Here we aim to quantitatively describe one aspect of human interaction with the goal to implement the results in the robotic system presented in [2] (see figure 1). Until now, several studies were done on human-robot interaction [1],[3], [4], [9]-[15] but only few of them [3],[9],[10],[12] were based on the analysis of human-human experiments. Our previous investigations on a simple hand-over scenario [9] demonstrated that human-robot interaction can indeed be improved significantly by implementing simple modifications to the robot behaviour such as human-like motion profiles following a minimum-jerk trajectory. Here, we extend our investigation to the approach phase preceding the hand-over. The case of a robot approaching a person for interaction has been investigated previously by others ([4],[11],[13],[15]), but the main parameters for the interaction, such as interpersonal distance, were usually taken from results presented by the anthropologist T.E. Hall [6] in the 1960's. According to Hall, the distance between humans in an interaction depends on their emotional relationship and cultural background. In order

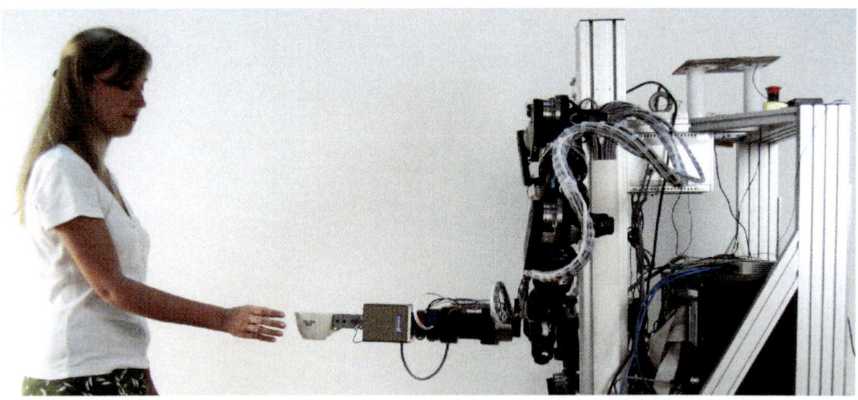

Fig. 1 Human-robot interaction scenario

to test these findings and to possibly define better human approach criteria, we investigated two experimental setups. In the first setup, one person, the transferring subject, walked towards a standing person, the receiving subject, to hand over an object. In the second setup, the receiving subject was replaced by a table.

2 Methods

2.1 Experimental Setup

We measured hand movements in human subjects during an approach and hand-over task using a motion tracking system (IS-600 Mark 2, InterSense Inc., USA), which tracks the position of wireless sensors using infrared and ultrasound signals. The size of the tracking area was $3m$ x $6m$ in the middle of a room of $38m^2$. We tested 26 pairs of subjects (6 female, 20 male, all right-handed). Each pair of subjects performed 4 consecutive trials. The item to be handed-over was a half full 0.5L plastic bottle with a height of 17 cm and a diameter of 10 cm with a weight of 210 g. Each subject served once as a standing person (receiving subject, R), and once as an approaching person (transporting subject, T). R and T were placed on predefined positions with a distance of 4.2 metres between them. R was instructed to stand on her position aware that T will give her an object. T was told to walk from the specified starting position towards R in order to deliver the bottle. Neither the speed nor the approaching path towards R was specified in order not to affect the natural behavior of T. The head position and orientation of the approaching person T was recorded by the tracking system by placing a 6-DOF sensor on a helmet carried on T's head. An additional tracking sensor, representing T's hand position, was placed on the item which was handed over (see figure 2). The head position of T was sampled with 150 Hz, the hand/object position with 20-50 Hz due to technical limitations of the tracking system. The recording of the data during each trial started with T already holding the object in his hand and finished when both R's and T's hands were on the object.

In a second setup the standing person R was replaced by a table. Here, 24 subjects were tested (6 female and 18 male, all right-handed). All subjects also participated in the first experiment. In this second scenario, T had to place the object on the table, which had a height of 1.2-1.3m that was individually derived from the first experiment in order to match the height of the hand-over in the first experiment. The table was placed in the same position as R in the first experiment. T received the instructions to go from the same starting position, as in the previous experiment, towards the table and to put the item on a specified position on the table. A tabletop of about $0.1\text{x}0.1m^2$ assured a well-defined position for placing the object (the same bottle as in experiment 1).

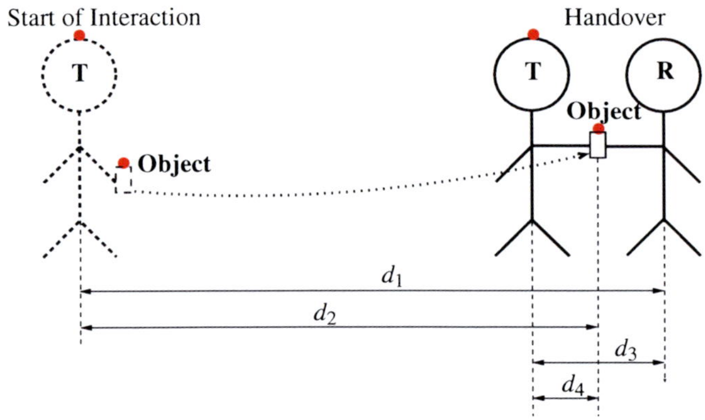

Fig. 2 Approach and handover scenario. The red circles denote the location of the tracking sensors. The depicted distances $d_1, \cdots, d_4$ are described in table 2 in section 3 for the human-human interaction scenario (left column).

2.2 Data Analysis

The approach and hand-over interaction was analysed off-line by deriving several parameters from the measured raw data using Matlab (The Mathworks, Natick, MA). The time and interpersonal distance of the start of the interaction was defined as the start of the hand movement in subject-coincident co-ordinates by employing a rule-based approach. The start of the interaction was determined as follows: 1) search backward, starting with the handover, for a maximum of T's vertical hand velocity, 2) from there, search backward for the first minimum in T's vertical hand position. This minimum was considered as start of interaction. Usually, it coincided with the time when the distance between T's head and T's hand started to increase rapidly (see figure 3).

Furthermore, the distance between T and his hand at the handover, the distance between R and T's hand at the handover, the interpersonal distance between T and R at the handover, and T's head and hand speed at the handover were determined. The same parameters were derived for the second scenario in which R had been replaced by a table. For statistical analysis, we used F-tests to compare variability, paired t-tests to compare means, and unpaired t-tests to compare against zero. A significance level of 0.05 was adopted in all tests.

3 Results

Figure 4 shows two representative examples of head and hand trajectories in top view. Since both subjects were facing each other at the start of each trial, the trajectories recorded for the approaching subject T are close to straight lines [7]. The

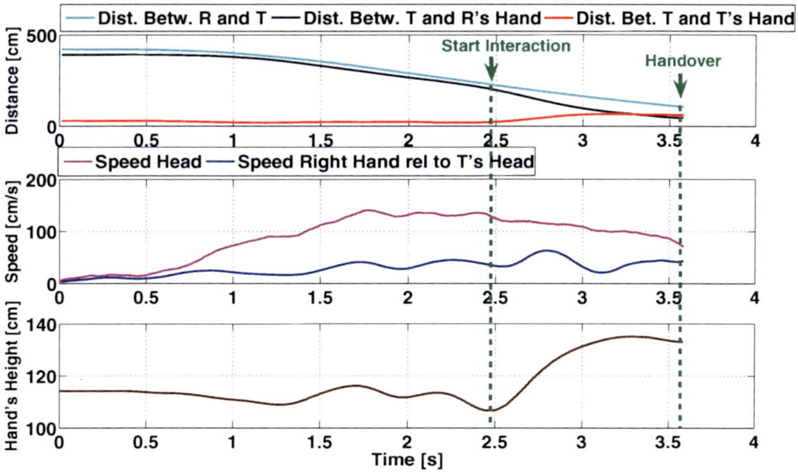

Fig. 3 Definition of the interaction phase

sinusoidal deviations of the head are caused by the human locomotor pattern, which leads to lateral head deviation with each step.

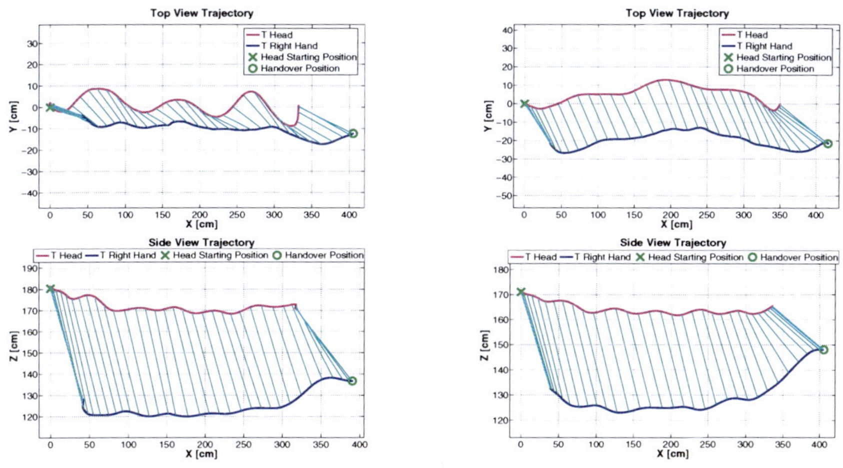

Fig. 4 Two examples of T's head (magenta) and hand (blue) trajectory for the approach and hand-over. Movement direction is from left to right. Corresponding points in time are connected by lines (cyan).

The results depicted in figures 5 and 6 show that the hand-over position is located closely to the midpoint between T and R, as found previously for hand-over between sitting subjects [8], although shifted slightly but significantly (t-test, $p<0.05$) to the right side (all the probands were right-handed). When expressed in interaction-related coordinates (right sides of figures 5 and 6), a significant (F-test, $p<0.05$) decrease of the variability of the X coordinate of the hand-over position was observed (see also table 1).The interaction-related coordinates are computed by re-expressing the handover position with respect to the mid-position between T and R for each trial.

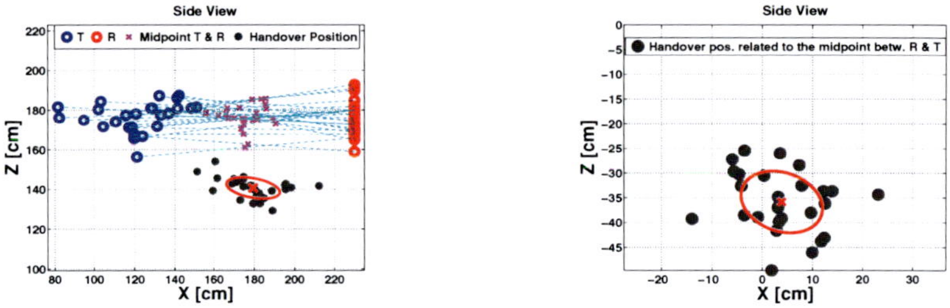

Fig. 5 Left: side view (XZ) of handover positions; the blue circles represent the average position of the head markers of the 26 transferring (T) subjects, the red circles those of the receiving (R) subjects. The purple crosses indicate the mid points between T and R, and the grey dots the handover positions. Right: handover positions redrawn relative to the midpoint between R and T as new origin.

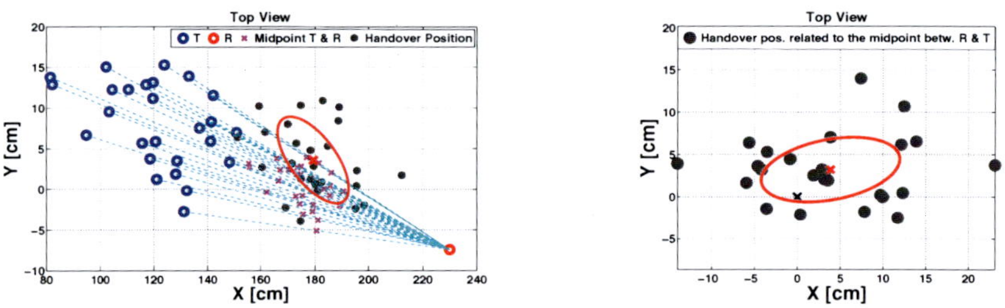

Fig. 6 Top view (XY) of handover positions, same data and same presentation as in figure 5. As seen in the right plot, the lateral position of the handover deviates slightly but significantly to the right side of the receiving subject. Note different axis scaling.

Table 1 Average and standard deviation of the hand-over position in a global and in an interaction-related coordinate system (N = 26 subjects))

	Global coordinates [cm]	Interaction-related coordinates [cm]
Mean	X=179.38 Y=3.6 Z=140.78	X=3.88 Y=3.23 Z=-35.77
Standard Deviation	X=13.14* Y=5.33 Z=5.37	X=8.09* Y=3.82 Z=6.22

* Variance significantly different (F-test).

Figure 7 shows examples of velocity profiles together with the corresponding height over ground for head and hand. The sinusoidal motion reflects the human locomotor pattern. Head velocity follows a bell-shaped curve: the experiment starts when T is still standing, then the speed increases while T approaches R, and finally the speed decreases until the handover. However, even at the actual handover (coinciding with the stop of the measurement, see 2.1), T's head and hand are still moving with a moderate speed. That means, the approaching person stops walking *after* the standing person has taken over the object. At the handover, T's head and T's hand had an average speed of 30 cm/s and 12.5 cm/s, respectively.

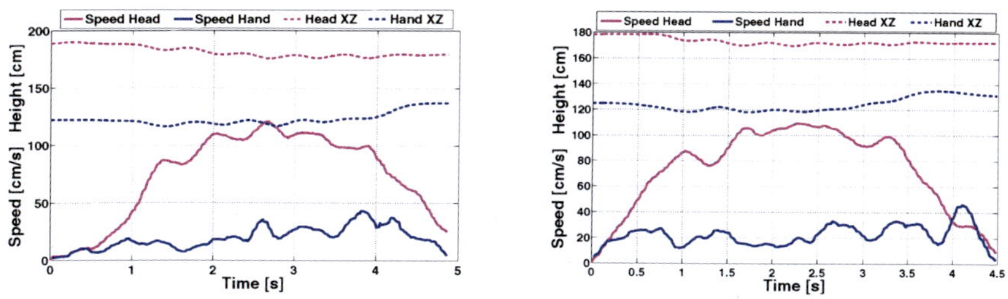

Fig. 7 Examples of velocity profiles and corresponding z-position (height) of head and hand. Hand velocity is expressed with respect to the head.

The results of the second scenario, in which the receiving subject R was replaced by a table, revealed very similar results. Table 2 compares the average values (±SD) of the most relevant parameters for the human-human interaction scenario (left column and figure 2) and the table scenario (right column). For the comparison, data were expressed relative to the handover position for experiment 1 and relative to the table position in experiment 2. A significantly different result was only found for the average peak hand velocity, which was faster in the interaction scenario.

Table 2 Average parameters (±SD) from the two experiments. Left: Interaction scenario; Right: Table scenario. d_1 and d_3 do not occur in the table scenario but these are still relevant for robotic applications. See figure 2 in section 2.1.

Parameter	Interaction (N = 26)	Table (N = 24)
d_1: Distance of T-R at start of handover [cm]	216,768±41.95	
d_2: Distance between T at handover initiation and handover position [cm]	160.6±22.45	157.13±25.97
d_3: Interpersonal distance at handover [cm]	116.02±19.15	
d_4: Distance between T and object at handover [cm]	64.68±8.73	69.19±13.27
Time between initiation and handover [s]	1.24±0.28	1.27±0.28
Peak head velocity of T [cm/s]	128.91±8.96	133.71±13.55
Peak hand velocity of T (relative to head) [cm/s]	84.8±22.56 *	114.71±35.95 *

* significantly different values for both scenarios ($p<0.05$).

The similarity of both experiments became even more evident when considering not only the average values, but comparing subject by subject. Figure 8 shows that the distance between T and the object at handover correlates significantly with the distance between T and the table at the moment of placing the object. Similarly, the distance at which the action is initiated, that is, the lifting of the hand for handover or placing starts, is correlated significantly in both scenarios.

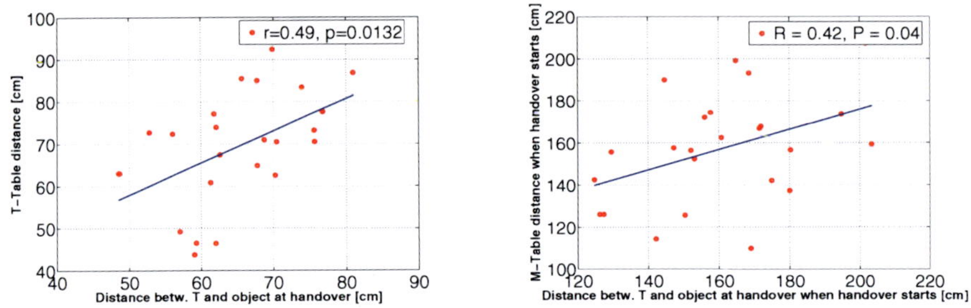

Fig. 8 Correlations between comparable parameters (see also table 2) in both scenarios. Dots represent values from one subject. X-axes: data for handover interaction from T to R, y-axis: data for placing of an object on a table by T.

We further analysed whether the final distance between both subject depended on the height and/or armlength of T and/or R. However, no significant correlations between the interpersonal distance at the handover and the height or armlength of the subjects were found.

3.1 Discussion

The present results show that approach and handover are smooth and dynamic actions with the different parts blending into each other. For example, the handover is initiated by lifting the object well in advance of the actual object transfer. The transfer itself happens while the transferring person is still moving, and, therefore, the approaching person T stops walking after the standing person R has grasped the object. The object transfer occurs at the midpoint between both subjects, thus confirming our earlier results for handover between two sitting subjects [8].

Our second experiment, in which the receiving person was replaced by a table, showed that not only the approach phase for handover or placing the object were very similar, but that even parameters such as distance between head and hand during handover are correlated with the respective parameter while placing the object on a table. That suggests that the large inter-individual differences for certain parameters did not depend on the second, receiving person being present, but were inter-individual preferences, which only depend on the transferring person. While previous work by other groups, e.g. [11], used results from the work of Hall [6] on social interaction to determine factors such as distance between subjects during interaction, our results suggest that, at least in a socially homogeneous group of subjects such as the one used here, additional social factors play a negligible role and the inter-individual variability in hand-over distance represent individual preferences for action in general. The only parameter which turned out to be significantly different between handover and placing an object was the peak hand velocity of the transferring subject, which was smaller for handover (see Table 2). We assume that during handover, a more accurate placement is required than when placing the object on the stationary table. This is in accordance with Fitts' law, which predicts a slower peak velocity for the smaller target [16].

However, even though the motion of the transferring subject is astonishingly independent of a receiving subject being present, the smooth progress of the handover from early lifting to actual transfer makes it easy for the receiving subject to infer the intention of the transferring partner. Moreover, the final position of the handover is determined by the end position of the transferring subject, and can, due to the smooth velocity trajectory during deceleration, be well predicted by the receiver. Even though we have so far only tested frontal approach, other work [7] has convincingly shown that goal-directed locomotor trajectories follow a maximal-smoothness constraint. Thus, as in the handover itself, where smooth hand trajectories are of importance for intention recognition [9], a smooth locomotor trajectory may be critical for seamless approach and hand-over.

Finally, we would like to point out that several of the present results are directly transferable to the robot. For instance, one can implement in the robot the distance to the receiver when the interaction starts and in addition the handover position. Yet the most important point is possibly that the approach, the start of the interaction by lifting the object appropriately, and the final handover are all performed *smoothly* and, in part, in *parallel* instead of being separate and successive motor programs.

Acknowledgements. This work is supported by the DFG cluster of excellence "CoTeSys" (`www.cotesys.org`). Furthermore, we would like to thank our research partner M.Sc. Omiros Kourakos from the Institute of Automatic Control Engineering, Technische Universität München for his support.

References

1. Alami, R., Clodic, A., Montreuil, V., Sisbot, E.A., Chatila, R.: Toward Human-Aware Robot Task Planning. In: Proceedings of AAAI Spring Symposium: To boldly go where no human-robot team has gone before, Stanford, USA (2006)
2. Althoff, D., Kourakos, O., Lawitzky, M., Mörtl, A., Rambow, M., Rohrmüller, F., Bršćić, D., Wollherr, D., Hirche, S., Buss, M.: A flexible architecture for real-time control in multi-robot systems. In: Ritter, H., et al. (eds.) Human Centered Robot Systems. COSMOS, vol. 6. Springer, Heidelberg (2009)

3. Balasuriya, J.C., Watanabe, K., Pallegedara, A.: ANFIS based active personal space for autonomous robots in ubiquitous environments. In: International Conference on Industrial and Information Systems, pp. 523–528 (2007)
4. Balasuriya, J.C., Watanabe, K., Pallegedara, A.: Giving robots some feelings towards interaction with humans in ubiquitous environment. In: International Conference on Industrial and Information Systems, pp. 529–534 (2007)
5. Fong, T., Nourbakhsh, I., Dautenhahn, K.: A survey of socially interactive robots. Robotics and Autonomous Systems, 143–166 (2003)
6. Hall, E.T.: The Hidden Dimension. Doubleday, New York (1966)
7. Hicheur, H., Pham, Q.-C., Arechavaleta, G., Laumond, J.-P., Berthoz, A.: The formation of trajectories during goal-oriented locomotion in humans. I. A stereotyped behaviour. European Journal of Neuroscience 26, 2376–2390 (2007)
8. Huber, M., Knoll, A., Brandt, T., Glasauer, S.: Handing-over a cube: spatial features of physical joint action. Annals of the New York Academy of Sciences 1164, 380–382 (2009)
9. Huber, M., Rickert, M., Knoll, A., Brandt, T., Glasauer, S.: Human-robot interaction in handing-over tasks. In: 7th IEEE International Symposium on Robot and Human Interactive Communication, pp. 107–112 (2008)
10. Kajikawa, S., Ishikawa, E.: Trajectory planning for hand-over between human and robot. In: 9th IEEE International Workshop on Robot and Human Interactive Communication, pp. 281–287 (2000)
11. Koay, K.L., Sisbot, E.A., Syrdal, D.S., Walters, M.L., Dautenhahn, K., Alami, R.: Exploratory Study of a Robot Approaching a Person in the Context of Handing Over an Object. In: Proceedings of AAAI Spring Symposium: Multidisciplinary Collaboration for Socially Assistive Robotics, AAAI Technical Report, pp. 18–24 (2007)
12. Nakauchi, Y., Simmons, R.: A social robot that stands in line. Autonomous Robots 12(3), 313–324 (2002)
13. Satake, S., Kanda, T., Glas, D.F., Imai, M., Ishiguro, H., Hagita, N.: How to approach humans? Strategies for social robots to initiate interaction. In: ACM/IEEE International Conference on Human-Robot Interaction, pp. 109–116 (2009)
14. Sisbot, E.A., Clodic, A., Alami, R., Ransan, M.: Supervision and motion planning for a mobile manipulator interacting with humans. In: ACM/IEEE International Conference on Human-Robot Interaction, pp. 327–334 (2008)
15. Walters, M.L., Dautenhahn, K., Woods, S., Koay, K.L.: Robotic etiquette: results from user studies involving a fetch and carry task. In: ACM/IEEE International Conference on Human-Robot Interaction, pp. 317–324 (2007)
16. Harris, C.M., Wolpert, D.M.: Signal-dependent noise determines motor planning. Nature 394, 780–784 (1998)

Modeling of Biomechanical Parameters Based on LTM Structures

Christoph Schütz, Timo Klein-Soetebier, and Thomas Schack

Abstract. Previous studies concerned with the interaction of cognition and biomechanics demonstrated correlations between 'global' parameters of a movement (e. g. duration) and the cognitive representation structure in long term memory. We asked if more 'local' biomechanical parameters (i. e. postures) are integrated into such a representation structure as well. To this end, the movement kinematics and representation structures of table tennis experts were measured for the forehand backspin serve and combined in a multilinear regression model. Results show that a few selected parameters of the ball's flight can be predicted with good accuracy, while task-relevant posture parameters cannot. Thus, the measurement of cognitive representation structures cannot be used for the joint angle modeling of movement kinematics.

1 Introduction

Each interaction with a three-dimensional environment requires a series of transformations between sensory and motor coordinate systems. Several of these transformations involve one-to-many mappings, which, in theory, create an infinite number of possible movement kinematics [7]. Experimental observations have demonstrated

Christoph Schütz · Timo Klein-Soetebier
Bielefeld University, PO 100 131, 33501 Bielefeld
e-mail: {christoph.schuetz,timo.klein-soetebier}@uni-bielefeld.de

Thomas Schack
Cognitive Interaction Technology, Center of Excellence, Bielefeld University, 33615 Bielefeld
CoR-Lab, Research Institute for Cognition and Robotics, Bielefeld University, 33615 Bielefeld
Bielefeld University, PO 100 131, 33501 Bielefeld
e-mail: thomas.schack@uni-bielefeld.de

H. Ritter et al. (Eds.): Human Centered Robot Systems, COSMOS 6, pp. 161–171.
springerlink.com

that, for a reasonably large class of movements, a number of kinematical parameters tend to remain invariant [5][3]. To create such a reproducible behaviour, the central nervous system has to reduce the redundant degrees of freedom that occur from the neural signal to movement kinematics [1].

Optimisation theory [7] and motor control strategies [18] provide means to impose constraints onto the movement selection system. Alternative approaches, which focus on the output part of the motor system, are the coupling of multiple degrees of freedom into 'functional synergies' [11][12] and the 'equilibrium point' hypothesis [10]. Latash states that only the terminal posture of a movement is imposed onto the musculoskeletal system, which then converges towards this set 'equilibrium point' based on feedback [10]. From a cognitive point of view, the representation of the terminal posture of a movement is much simpler than the representation and control of the complete movement dynamics [6][17].

According to Schack [19], such movement representations are organised and stored in long term memory (LTM) by way of their anticipated effects. Output patterns of motor commands are triggered with regard to the cognitive reference structure. Bernstein [1] and Hoffmann [4] hypothesised that concepts are of particular significance for the formation of such perceptual-cognitive motor representations. With reference to this work, Schack [22][21] proposed that representation structures are organised as hierarchical frameworks of 'basic action concepts' (BACs). Each BAC integrates functional and sensory features of a movement event [25]. Its position in the multidimensional feature space is measured with the structural dimensional analysis-motoric (SDA-M [9][20]). Based on the Euclidean distances between the BACs in feature space, a hierarchical representation of the LTM structure can be created.

The representation structures serve as cognitive references for the creation of motor patterns [19]. In order to realise anticipated movement effects, the motor system has to possess an inverse model of the movement [8]. Therefore, some biomechanical parameters of the movement have to be embedded into the representation structure [27]. Jeannerod [6] and Rosenbaum [17] state that at least the terminal posture of a movement has to be represented. A previous study by Schack [19] demonstrated correlations between biomechanical parameters of a movement and the according LTM structures. One shortcoming of the study was the fact that only 'global' movement parameters (e. g. duration) were investigated. We asked if more 'local' biomechanical parameters (i. e. postures) are also integrated into the representation structure of complex movements.

To address this question, we measured movement kinematics and representation structures of expert table tennis players. Previous studies in table tennis demonstrated performance and accuracy differences between experts and novices [16] and analysed activity structures of expert table tennis players during a match [23]. To the best of our knowledge, the LTM structures of table tennis players have not been investigated. The forehand backspin serve, used in the present study, is well suited for a lab environment. It is similar in complexity to the tennis serve, that has been successfully analysed with the SDA-M before [21].

2 Experimental Design

2.1 Participants

Nine expert table tennis players (male, average age 22.5 years, average expertise 9.1 years) participated in the experiment. Eight participants were right handed and one was left handed (self-report). Participants characterised themselves as neurologically healthy and were nave to the purpose of the study. Before the experiment, each participant provided his informed consent and read a detailed set of instructions concerning the required task. In addition, the LTM structure of a high-level expert (male, age 26 years, expertise 20 years, 2nd Bundesliga) was measured as a reference structure.

2.2 Biomechanical Analysis

A. Experimental procedure

Twenty spherical retro reflective markers (diameter 14 mm) were attached to bony landmarks on the arms and thorax of the participant (see Table 1). Three additional markers were attached to the racket, defining the racket plane. The table tennis balls (38 mm diameter) were coated with retro reflective material, resulting in a ball diameter of 40 mm.

Movement data was recorded using an optical motion capture system (Vicon Motion Systems, Oxford, UK) consisting of 6 MX-3+ CCD cameras with 200 Hz temporal and approximately 0.5 mm of spatial resolution. Cartesian coordinates of the markers and the ball were calculated from the camera data via triangulation. Marker trajectories were manually labeled in Vicon Nexus 1.1 and exported to a *c3d*-file for post processing.

Two calibration movements for each arm were recorded, one in the horizontal plane and one in the frontal plane. The participant was positioned in front of a custom table tennis table (2.74 m long, 1.525 m wide and 0.76 m high with a net height of 15.25 cm). Each trial began from an initial T-pose, with both arms extended. The racket was positioned on the table in front of the participant. On a signal from the experimenter, the participant (1) reached for the racket with his dominant hand, (2) received a ball from a research assistant, (3) performed a forehand backspin serve, (4) placed the racket back onto the table and (5) returned to the T-pose. The quality of the serve was assessed by a neutral table tennis expert based on four criteria: spin, velocity, height and placement of the ball. The sequence was repeated until a total of 50 serves was reached. The entire experiment lasted approximately 60 min. Participants were debriefed after the experiment.

Table 1 Real marker positions (anatomical landmarks), virtual marker positions (calculated), segment definitions and Euler angle definitions of the kinematical model

	Real marker positions			
	anatomical landmark			**anatomical landmark**
PX	*Processus xiphoideus*		ACR	*Acromion*
IJ	*Incisura jugularis*		ELM	medial epicondyle
C7	7th cervical vetrebra		ELL	lateral epicondyle
T8	8th thoracal vetrebra		WRU	ulnar styloid
AI	*Angulus inferior*		WRR	radial styloid
AS	*Angulus superior*		MCP	*Os metacarpale tertium*[a]
	Virtual marker positions			
	calculation			**calculation**
SJC	*shoulder centre:* sphere fit[b]		THOU	$(C7+IJ)/2$
EJC	$(ELM+ELL)/2$		THOL	$(T8+PX)/2$
WJC	$(WRU+WRR)/2$		THOF	$(IJ+PX)/2$
HC	*hand centre:* trigonometry[b]		THOB	$(C7+T8)/2$
	Segment definitions			
	x-axis	**y-axis**	**z-axis**	**r (support vector)**
Thorax	$Y \times Z$	$THOU-THOL$	$\mathbf{r} \times Y$	$THOF-THOB$
Upper arm	$Y \times Z$	$SJC-EJC$	$\mathbf{r} \times Y$	$WJC-EJC$
Lower arm	$\mathbf{r} \times Y$	$EJC-WJC$	$X \times Y$	$WRU-WRR$
Hand	$\mathbf{r} \times Y$	$WJC-HC$	$X \times Y$	$WRU-WRR$
	Euler angle definitions			
	'Parent'	**'Child'**	**Euler sequence (floating axes)**	
Shoulder	Thorax	Upper arm	$y-x-y$	
Elbow	Upper arm	Lower arm	$z-x-y$	
Wrist	Lower arm	Hand	$z-x-y$	

[a] dorsal side of the *capitulum*. [b] refer to section 2.2 B.

B. Kinematical calculations

Kinematical calculations were conducted in MATLAB (R2008a, The MathWorks, Natick, MA). From the set of 50 forehand backspin serves performed by each participant, the 5 highest rated serves were included in the final analysis.

Wrist joint centres (*WJC*) were calculated halfway between the radial and the ulnar styloid marker, elbow joint centres (*EJC*) halfway between the lateral and medial epicondyle marker of the respective arm (see Table 1). Based on the two calibration movements, shoulder joint centres (*SJC*) were calculated in a local scapula coordinate system via sphere fitting, using least-squares optimisation [14][15]. The local positions were re-transformed to global positions of the shoulder joint centres for each movement. Two direction vectors for the hand were calculated ($V_1 = MCP - WJC$, $V_2 = WRU - WRS$). The hand centre (*HC*) was defined on a plane normal to $V_1 \times (V_2 \times V_1)$, positioned palmar from *MCP* at a distance of $(handthickness + markerdiameter)/2$.

Segment definitions were based on the ISB recommendations on the definition of joint coordinate systems for the upper body [26]. Joint angles were calculated via Euler angles between adjoining segments (see Table 1). To standardise Euler angles for the left and right arm, markers on the left side of the body were mirrored along the thorax plane before calculation. Four joint angles were measured for the dominant arm: (1) shoulder forward/backward rotation (2) elbow flexion/extension (3) elbow pronation/supination and (4) wrist adduction/abduction. Joint angle values were analysed at four discrete points in time, corresponding to (1) the initiation of the ball toss (initial posture) (2) the maximum retraction of the arm, (3) the moment of ball racket contact and (4) the moment of ball table contact (final posture). For the analysis of intrasubject posture variability, participants' variance of the joint angles was calculated for five serves.

The time lag between (1) and (3) was used to define the duration of the movement. Four ball parameters were calculated, (1) the initial velocity (2) the direction (3) the spin and (4) the amount of side spin.

2.3 LTM Structure Analysis

The forehand backspin serve was subdivided into 13 BACs based on literature and evaluation by coaches: (1) *legs a little more than shoulder width apart*, (2) *toss ball to head height*, (3) *shift centre of mass backward*, (4) *move racket backward*, (5) *rotate hip and shoulder to the right (left)*, (6) *lay the wrist back*, (7) *focus on the ball*, (8) *lower body towards the point of ball contact*, (9) *rotate hip and shoulder into the ball*, (10) *move racket downward and forward*, (11) *open racket*, (12) *chopping motion of the wrist* and (13) *sweep racket through*.

Measurement of the cognitive structure was based on the SDA-M [21]. The participant was asked to perform a hierarchical split procedure on a laptop. A randomly selected BAC was presented as an anchoring unit on top of a randomly ordered list of the remaining BACs. The participant was asked to sort each of the remaining BACs into a positive or a negative subset, depending on whether or not they were 'functionally related' to the anchoring unit. The subsets were submitted to the same split procedure until no further subdivision was possible. As every BAC was used as an anchoring unit, the procedure resulted in 13 decision trees per participant. For each BAC within a decision tree, the sum of its nodes was calculated. The normalised sum vector of each anchoring BAC defined its position in the multidimensional feature space. Based on the positions of the BACs, a Euclidean distance matrix was calculated. Distances were submitted to a hierarchical cluster analysis (unweighted average linkage) to create a represention of the LTM structure. A significance level of $p = .05$ was chosen for all analyses.

2.4 Combination of Biomechanical Parameters and LTM Structures

Each of the 21 biomechanical parameters (cf. section 2.2B) was used as the response variable of a multilinear regression model. A subset of plausible BACs for the prediction of the parameter was selected. Within the subset, all possible combinations of two distances d_1 and d_2 between the BACs in LTM (see Figure 1A) were tested as dependent variables of the corresponding multilinear regression model. For example, a single distance combination would be d_1 = distance between the BACs (4) *move racket backward* and (10) *move racket downward and forward* and d_2 = distance between the BACs (10) and (12) *chopping motion of the wrist*. Non-significant correlations were discarded, significant correlations ($p < .05$) were re-evaluated for the interpretability of the selected distances with regards to content.

3 Results

Figure 1A represents the LTM structure of the high-level expert (2nd Bundesliga) for the forehand backspin serve. The movement is separated into two distinct movement phases. The pre-activation phase consists of the BACs (2) *toss ball to head height*, (3) *shift centre of mass backward*, (4) *move racket backward*, (5) *rotate hip and shoulder to the right (left)*, (6) *lay the wrist back* and (11) *open racket*. The strike phase includes the BACs (8) *lower body towards the point of ball contact*, (9) *rotate hip and shoulder into the ball*, (10) *move racket downward and forward*,

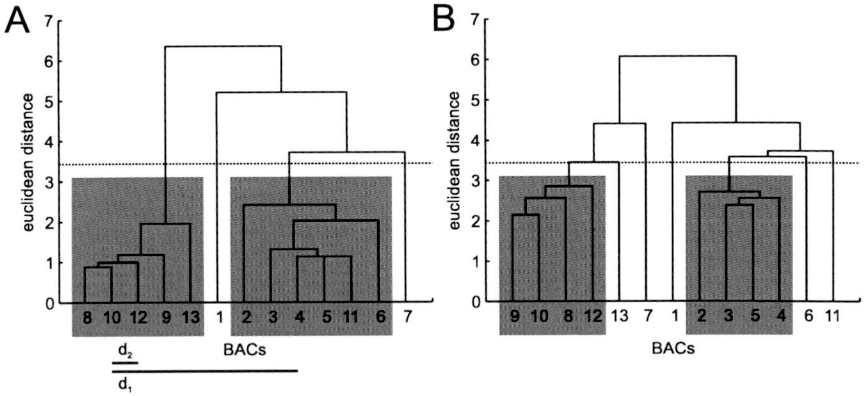

Fig. 1 [A] Hierarchical LTM structure of the high-level expert (2nd Bundesliga). The dotted line depicts the critical distance ($p = .05$). The movement is separated into two distinct movement phases (grey boxes), a pre-activation and a strike phase. A plausible combination of two distances (d_1, d_2) between BACs is depicted. [B] Mean LTM structure of the participant group. A similar separation into two movement phases is found.

(12) *chopping motion of the wrist* and (13) *sweep racket through*. Two BACs have not been assigned to a movement phase, namely (1) *legs a little more than shoulder width apart* and (7) *focus on the ball*.

Figure 1B represents the mean LTM structure of the table tennis players in our study. The hierarchical structure is similar to the expert structure, with a separation into two principal movement phases. A smaller number of BACs is assigned to each of the two phases (based on the significance level of $p = .05$). The pre-activation phase includes the elements 2, 3, 4 and 5, while the strike phase consists of the elements 8, 9, 10 and 12. The cluster solution differs significantly from the cluster solution of the high-level expert ($\lambda = 0.65$, $\lambda_{krit} = 0.68$).

The multilinear regression model reveals significant correlations for three ball specific parameters, as well as for the movement duration. Ball parameters that can be predicted based on the LTM structure are (1) the spin ($d_1 = 10-11, d_2 = 11-12$, $R^2 = 0.88, p < .01$), (2) the amount of side spin ($d_1 = 8-9, d_2 = 10-11, R^2 = 0.85$, $p < .01$) and (3) the direction of the ball ($d_1 = 4-10$, $d_2 = 9-12$, $R^2 = 0.96$, $p < .001$). Movement duration can be predicted with reasonable accuracy ($d_1 = 4-10$, $d_2 = 10-12$, $R^2 = 0.95$, $p < .001$). Examples for ball direction and movement duration are shown in Figure 2A and 2B. No significant correlations between any of the posture parameters and the LTM structure were found.

As the correlation analysis yielded no results for the posture parameters, we investigated intrasubject variance of the posture for three different points in time: (1) the moment of maximum retraction of the arm (before ball contact) (2) the moment of ball racket contact and (3) the moment of first ball table contact (after ball contact). The analysis shows similar joint angle variance before and during contact, but significantly increased joint angle variance after the contact (elbow flexion: $t(8) = -3.83$, $p < .01$; elbow rotation: $t(8) = -2.34$, $p < .05$; wrist adduction/abduction: $t(8) = -3.43$, $p < .01$). Examples for elbow flexion and rotation are shown in Figure 3A and 3B.

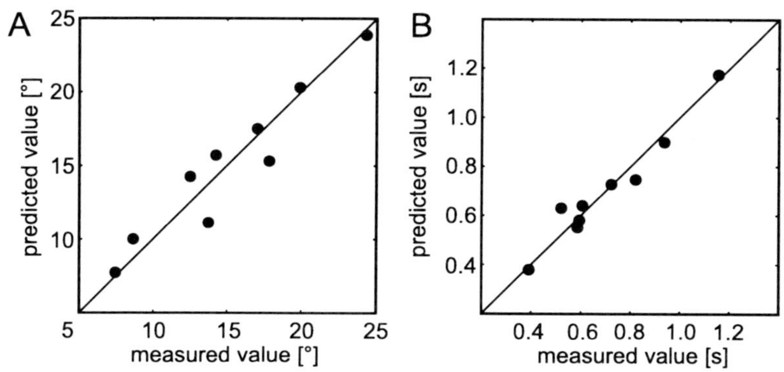

Fig. 2 Result of the multilinear regression model. [A] Prediction of ball direction based on LTM structures ($d_1 = 4-10$, $d_2 = 9-12$, $R^2 = 0.96$, $p < .001$) [B] Prediction of movement duration based on LTM structures ($d_1 = 4-10$, $d_2 = 10-12$, $R^2 = 0.95$, $p < .001$).

4 Discussion

Previous studies concerned with the interaction of cognition and biomechanics demonstrated correlations between 'global' parameters of a movement and its representation in LTM. We asked if more 'local' biomechanical parameters (i. e. postures) are integrated into the representation structure as well. To this end, we employed the SDA-M [21] to measure the LTM structures of expert table tennis players.

The result of the hierarchical cluster analysis for the high-level expert reveals a well-structured movement representation, which is separated into two distinct movement phases, pre-activation and strike phase (see Figure 1A). The representation structure matches the movement structure, as described by Schack [19]. The cluster solution of the participant group (see Figure 1B) reveals a less defined phase structure, which differs significantly ($p < .05$) from the high-level expert structure. Schack and Mechsner [21] demonstrated that movement-related structures of expert tennis players are better organized and adapted to biomechanical and functional demands of the movement than those of novices. A more detailed study by Bläsing and colleagues [2] has shown that, with increasing skill level, the representation gradually converges towards the expert structure. The similarity of the overall hierarchical structure of both the high-level expert and the participant group thus indicates a reasonably good level of expertise of the participants.

The multilinear regression model can predict selected parameters of the ball's flight and the movement duration with reasonable accuracy. The BACs used for the predictor distances can be meaningfully interpreted. For example, movement duration (time lag between movement initiation and ball racket contact) is predicted based on the distances between (4) *move racket backward* and (10) *move racket downward and forward* and between (10) and (12) *chopping motion of the wrist.* One can assume that, from a temporal point of view, (4) is closely related to movement initiation and (12) is closely related to the moment of ball racket contact. Thus,

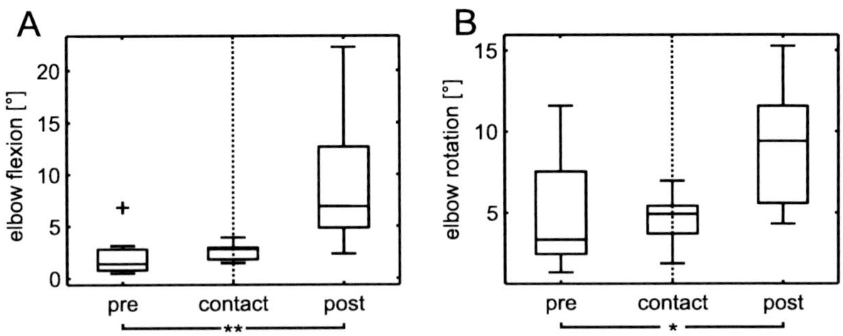

Fig. 3 Intrasubject joint angle variance measured before, during and after the moment of ball racket contact for [A] elbow flexion ($t(8) = -3.83$, $p < .01$) and [B] elbow rotation ($t(8) = -2.34$, $p < .05$)

the sequence of the BACs 4 – 10 – 12, as defined by the two predictor distances in LTM, spans the full time range of the movement.

Similar results concerning the duration of the movement have already been described by Schack [19]. The goal of the present study was to extend these previous results, which were restricted to more 'global' parameters of the movement like duration, to the analysis of joint angles. Yet none of the investigated posture parameters showed a significant correlation with the movement representation in LTM. According to Schack [19], actions are represented and stored in memory by way of their anticipated effects. Jeannerod [6] and Rosenbaum [17], on the other hand, state that at least the terminal posture of a movement has to be represented. Results indicate that, for the forehand backspin serve, only the achieved effect of the action on the ball is represented in the LTM structure, while the terminal posture is not.

One can assume that the terminal posture of the serve is irrelevant to the effect of the action and thus, does not influence the representation structure. To address this question, we calculated the intrasubject variance of the posture parameters before, during and after the moment of ball racket contact. Latash [11] claims that inherent properties of the musculoskeletal system create a certain amount of variability. According to the uncontrolled-manifold-hypothesis [13], the motor control system reduces the variability of task-relevant parameters of the movement [24] via increased variability of the task-irrelevant parameters ('functional variability'). The present results show that the variance of the posture parameters is equally low before and during the moment of ball racket contact, and increases significantly after the contact (see Figure 3). These results indicate that, for the table tennis serve, movement kinematics before and during the moment of ball racket contact are *not* task irrelevant and thus, should influence the representation structure.

In conclusion, the present study demonstrates that the representation structures measured via SDA-M can be used to predict and model selected parameters of a movement, providing a valuable tool for the creation of more human-like robotic systems. On the other hand, the method fails to reproduce any of the task-relevant posture parameters and, thus, the data at hand cannot be used for the joint angle modelling of movement kinematics.

References

1. Bernstein, N.A.: The co-ordination and regulation of movements, 1st english edn. Pergamon Press Ltd., Oxford (1967)
2. Blaesing, B., Tenenbaum, G., Schack, T.: The cognitive structure of movements in classical dance. Psychology of Sport and Exercise 10(3), 350–360 (2009)
3. Flash, T., Hogan, N.: The coordination of arm movements - an experimentally confirmed mathematical model. Journal of Neuroscience 5(7), 1688–1703 (1985)
4. Hoffmann, J.: Vorhersage und Erkenntnis. Hogrefe, Goettingen (1993)
5. Hogan, N.: An organizing principle for a class of voluntary movements. Journal of Neuroscience 4(11), 2745–2754 (1984)

6. Jeannerod, M.: Motor representations: One or many? Behavioral and Brain Sciences 19(4), 763 (1996)
7. Jordan, M., Wolpert, D.M.: Computational motor control. In: Gazzaniga, M. (ed.) The Cognitive Neurosciences. MIT Press, Cambridge (1999)
8. Kalveram, K.: The inverse problem in cognitive, perceptual and proprioceptive control of sensorimotor behaviour. Towards a biologically plausible model of the control of aiming movements. International Journal of Sport and Exercise Psychology 2, 255–273 (2004)
9. Lander, H.J., Lange, K.: Research on the structural and dimensional analysis of conceptually represented knowledge. Zeitschrift fuer Psychologie 204(1), 55–74 (1996)
10. Latash, M.L.: Control of multijoint reaching movement: The elastic membrane metapho. In: Latash, M.L. (ed.) Progress in Motor Control, vol. 1, pp. 315–360. Human Kinetics, Champaign (1998)
11. Latash, M.L.: Neurophysiological basis of movement. Human Kinetics, Champaign (1998)
12. Latash, M.L., Jaric, S.: Instruction-dependent muscle activation patterns within a two-joint-synergy: separating mechanics from neurophysiology. Journal of Motor Behaviour 30(3), 194–198 (1998)
13. Latash, M.L., Scholz, J.P., Schoener, G.: Motor control strategies revealed in the structure of motor variability. Exercise and Sport Sciences Reviews 30(1), 26–31 (2002)
14. Levenberg, K.: A method for the solution of certain problems in least squares. Quart. Appl. Math. 2, 164–168 (1944)
15. Marquardt, D.: An algorithm for least-squares estimation of nonlinear parameters. Journal of the Society for Industrial and Applied Mathematics 11(2), 431–441 (1963)
16. Rodrigues, S.T., Vickers, J.N., Williams, A.M.: Head, eye and arm coordination in table tennis. Journal of Sports Sciences 20(3), 187–200 (2002)
17. Rosenbaum, D.A., Cohen, R.G., Jax, S.A., Weiss, D.J., van der Wel, R.: The problem of serial order in behavior: Lashley's legacy. Human Movement Science 26(4), 525–554 (2007)
18. Rosenbaum, D.A., Marchak, F., Barnes, H.J., Vaughan, J., Slotta, J.D., Jorgensen, M.J.: Constraints for action selection - overhand versus underhand grips. Attention and Performance 13, 321–342 (1990)
19. Schack, T.: The relationship between motor representation and biomechanical parameters in complex movements: Towards an integrative perspective of movement science. European Journal of Sport Science 3(2), 1–13 (2003)
20. Schack, T.: The cognitive architecture of complex movement. International Journal of Sport and Exercise Psychology; Special Issue: The construction of action - new perspectives in movement science. Part II 2(4), 403–438 (2004)
21. Schack, T., Mechsner, F.: Representation of motor skills in human long-term memory. Neuroscience Letters 391(3), 77–81 (2006)
22. Schack, T., Tenenbaum, G.: Perceptual and cognitive control in action - a preface. International Journal of Sport and Exercise Psychology; Special Issue: The construction of action - new perspectives in movement science. Part I 2(3), 207–209 (2004)
23. Seve, C., Saury, J., Ria, L., Durand, M.: Structure of expert players' activity during competitive interaction in table tennis. Research Quarterly for Exercise and Sport 74(1), 71–83 (2003)

24. Shim, J.K., Latash, M.L., Zatsiorsky, V.M.: Prehension synergies: Trial-to-trial variability and principle of superposition during static prehension in three dimensions. J. Neurophysiol. 93(6), 3649–3658 (2005)
25. Verwey, W.B., Abrahamse, E.L., Jimenez, L.: Segmentation of short keying sequences does not spontaneously transfer to other sequences. Human Movement Science 28(3), 348–361 (2009)
26. Wu, G., van der Helm, F.C.T., Veeger, H.E.J., Makhsous, M., Van Roy, P., Anglin, C., Nagels, J., Karduna, A.R., McQuade, K., Wang, X.G., Werner, F.W., Buchholz, B.: ISB recommendation on definitions of joint coordinate systems of various joints for the reporting of human joint motion - part II: shoulder, elbow, wrist and hand. Journal of Biomechanics 38(5), 981–992 (2005)
27. Zatsiorsky, V.M.: Kinematics of human motion. Human Kinetics, Champaign (1998)

Towards Meaningful Robot Gesture

Maha Salem, Stefan Kopp, Ipke Wachsmuth, and Frank Joublin

Abstract. Humanoid robot companions that are intended to engage in natural and fluent human-robot interaction are supposed to combine speech with non-verbal modalities for comprehensible and believable behavior. We present an approach to enable the humanoid robot ASIMO to flexibly produce and synchronize speech and co-verbal gestures at run-time, while not being limited to a predefined repertoire of motor action. Since this research challenge has already been tackled in various ways within the domain of virtual conversational agents, we build upon the experience gained from the development of a speech and gesture production model used for our virtual human Max. Being one of the most sophisticated multi-modal schedulers, the Articulated Communicator Engine (ACE) has replaced the use of lexicons of canned behaviors with an on-the-spot production of flexibly planned behavior representations. As an underlying action generation architecture, we explain how ACE draws upon a tight, bi-directional coupling of ASIMO's perceptuo-motor system with multi-modal scheduling via both efferent control signals and afferent feedback.

Maha Salem
Research Institute for Cognition and Robotics, Bielefeld University, Germany
e-mail: msalem@cor-lab.uni-bielefeld.de

Stefan Kopp
Sociable Agents Group, Bielefeld University, Germany
e-mail: skopp@techfak.uni-bielefeld.de

Ipke Wachsmuth
Artificial Intelligence Group, Bielefeld University, Germany
e-mail: ipke@techfak.uni-bielefeld.de

Frank Joublin
Honda Research Institute Europe, Offenbach, Germany
e-mail: frank.joublin@honda-ri.de

H. Ritter et al. (Eds.): Human Centered Robot Systems, COSMOS 6, pp. 173–182.
springerlink.com

1 Introduction

Non-verbal expression via gesture is an important feature of social interaction, frequently used by human speakers to emphasize or supplement what they express in speech. For example, pointing to objects being referred to or giving spatial directions conveys information that can hardly be encoded solely by speech. Accordingly, humanoid robot companions that are intended to engage in natural and fluent human-robot interaction must be able to produce speech-accompanying non-verbal behaviors from conceptual, to-be-communicated information. Forming an integral part of human communication, hand and arm gestures are primary candidates for extending the communicative capabilities of social robots.

According to McNeill [13], co-verbal gestures are mostly generated unconsciously and are strongly connected to speech as part of an integrated utterance, yielding semantic, pragmatic and temporal synchrony between both modalities. This suggests that gestures are influenced by the communicative intent and by the accompanying verbal utterance in various ways. In contrast to task-oriented movements like reaching or grasping, human gestures are derived to some extent from a kind of internal representation of shape [8], especially when iconic or metaphoric gestures are used. Such characteristic shape and dynamical properties exhibited by gestural movement enable humans to distinguish them from subsidiary movements and to perceive them as meaningful [17]. Consequently, the generation of co-verbal gestures for artificial humanoid bodies, e.g., as provided for virtual agents or robots, demands a high degree of control and flexibility concerning shape and time properties of the gesture, while ensuring a natural appearance of the movement.

In this paper, we first discuss related work, highlighting the fact that not much research has so far focused on the generation of robot gesture (Section 2). In Section 3, we describe our multi-modal behavior realizer, the Articulated Communicator Engine (ACE), which implements the speech-gesture production model originally designed for the virtual agent Max and is now used for the humanoid robot ASIMO. We then present a concept for the generation of meaningful arm movements for the humanoid robot ASIMO based on ACE in Section 4. Finally, we conclude and give an outlook of future work in Section 5.

2 Related Work

At present, the generation together with the evaluation of the effects of robot gesture is largely unexplored. In traditional robotics, recognition rather than synthesis of gesture is mainly brought into focus. In existing cases of gesture synthesis, however, models typically denote object manipulation serving little or no communicative function. Furthermore, gesture generation is often based on prior recognition of perceived gestures, hence the aim is often to imitate these movements. In many cases in which robot gesture is actually generated with a communicative intent, these arm movements are not produced at run-time, but are pre-recorded for demonstration purposes and are not finely coordinated with speech. Generally, only a few approaches share any similarities with ours, however, they are mostly realized on less sophisticated platforms with less complex robot bodies (e.g., limited mobility, less degrees of freedom (DOF), etc.). One example is the personal robot Maggie [6] whose aim is to interact with humans in a natural way, so that a peer-to-peer relationship can be established. For this purpose, the robot is equipped with a set of pre-defined gestures, but it can also learn some gestures from the user. Another example of robot gesture is given by the penguin robot Mel [16] which is able to engage with humans in a collaborative conversation, using speech and gesture to indicate engagement behaviors. However, gestures used in this context are predefined in a set of action descriptions called the "recipe library". A further approach is that of the communication robot Fritz [1], using speech, facial expression, eye-gaze and gesture to appear livelier while interacting with people. Gestures produced during interactional conversations are generated on-line and mainly consist of human-like arm movements and pointing gestures performed with eyes, head, and arms.

As Minato et al. [14] state, not only the behavior but also the appearance of a robot influences human-robot interaction. Therefore, the importance of the robot's design should not be underestimated if used as a research platform to study the effect of robot gesture on humans. In general, only few scientific studies regarding the perception and acceptance of robot gesture have been carried out so far. Much research on the human perception of robots depending on their appearance, as based on different levels of embodiment, has been conducted by MacDorman and Ishiguro [12], the latter widely known as the inventor of several android robots. In their testing scenarios with androids, however, non-verbal expression via gesture and gaze was generally hard-coded and hence pre-defined. Nevertheless, MacDorman and Ishiguro consider androids a key testing ground for social, cognitive, and neuroscientific theories. They argue that they provide an experimental apparatus that can be controlled more precisely than any human actor. This is in line with initial results, indicating that only robots strongly resembling humans can elicit the broad spectrum of responses that people typically direct toward each other. These findings highlight the importance of the robot's design when used as a research platform for the evaluation of human-robot interaction scenarios.

While being a fairly new area in robotics, within the domain of virtual humanoid agents, the generation of speech-accompanying gesture has already been addressed in various ways. Cassell et al. introduced the REA system [2] over a decade ago, employing a conversational humanoid agent named Rea that plays the role of a real estate salesperson. A further approach, the BEAT (Behavior Expression Animation Toolkit) system [3], allows for appropriate and synchronized non-verbal behaviors by predicting the timing of gesture animations from synthesized speech in which the expressive phase coincides with the prominent syllable in speech. Gibet et al. generate and animate sign-language from script-like specifications, resulting in a simulation of fairly natural movement characteristics [4]. However, even in this domain most existing systems either neglect the meaning a gesture conveys, or they simplify matters by using lexicons of words and canned non-verbal behaviors in the form of pre-produced gestures.

In contrast, the framework underlying the virtual agent Max [9] is geared towards an integrated architecture in which the planning of both content and form across both modalities is coupled [7], hence giving credit to the meaning conveyed in non-verbal utterances. According to Reiter and Dale [15], computational approaches to generating multi-modal behavior can be modeled in terms of three consecutive tasks: firstly, determining *what* to convey (i.e., content planning); secondly, determining *how* to convey it (i.e., micro-planning); finally, realizing the planned behaviors (i.e., surface realization). Although the Articulated Communicator Engine (ACE) itself operates on the surface realization layer of the generation pipeline, the overall system used for Max also provides an integrated content planning and micro-planning framework [7]. Within the scope of this paper, however, only ACE is considered and described, since it marks the starting point required for the interface endowing the robot ASIMO with similar multi-modal behavior.

3 An Incremental Model of Speech-Gesture Production

Our approach is based on straightforward descriptions of the designated outer form of the to-be-communicated multi-modal utterances. For this purpose, we use MURML [11], the XML-based Multi-modal Utterance Representation Markup Language, to specify verbal utterances in combination with co-verbal gestures [9]. These, in turn, are explicitly described in terms of form features (i.e., the posture aspired for the gesture stroke), specifying their affiliation to dedicated linguistic elements based on matching time identifiers. Fig. 1 shows an example of a MURML

```
<definition><utterance>
 <specification>
 And now take the object <time id="t1" chunkborder="true"/>
 and make it <time id="t2"/> this big. <time id="t3"/>
 </specification>
 <behaviorspec>
  <gesture id="gesture_1" scope="hand">
   <affiliate onset="t2" end="t3" focus="this"/>
    <constraints>
     <symmetrical dominant="right_arm" symmetry="SymMS">
      <parallel>
       <static slot="HandShape" value="BSflat(FBround all o)"/>
       <static slot="ExtFingerOrientation" value="DirA"/>
       <static slot="PalmOrientation" value="DirL"/>
       <static slot="HandLocation" value="LocChest LocCenterRight LocNorm"/>
      </parallel>
     </symmetrical>
    </constraints>
  </gesture>
 </behaviorspec>
</utterance></definition>
```

Fig. 1 Example of a MURML specification for multi-modal utterances specification which can be used as input for our production model. For more information on MURML see [11].

The concept underlying the multi-modal production model is based on an empirically suggested assumption referred to as *segmentation hypothesis* [13], according to which the co-production of continuous speech and gesture is organized in successive segments. Each of these, in turn, represents a single idea unit which we refer to as a *chunk* of speech-gesture production. A given chunk consists of an intonation phrase and a co-expressive gesture phrase, concertedly conveying a prominent concept [10]. Within a chunk, synchrony is mainly achieved by gesture adaptation to structure and timing of speech, while absolute time information is obtained at phoneme level and used to establish timing constraints for co-verbal gestural movements. Given the MURML specification shown in Fig. 1, the correspondence between the verbal phrase and the accompanying gesture is established by the <time id="..."/> tag with unique identifier attributes. Accordingly, the beginning and ending of the affiliate gesture is defined using the <affiliate onset="..." end="..."/> tag. The incremental production of successive coherent chunks is realized by processing each chunk on a separate 'blackboard' running through a sequence of states (Fig. 2). These states augment the classical two-phase planning - execution procedure with additional phases, in which the production process of subsequent chunks can interact with one another.

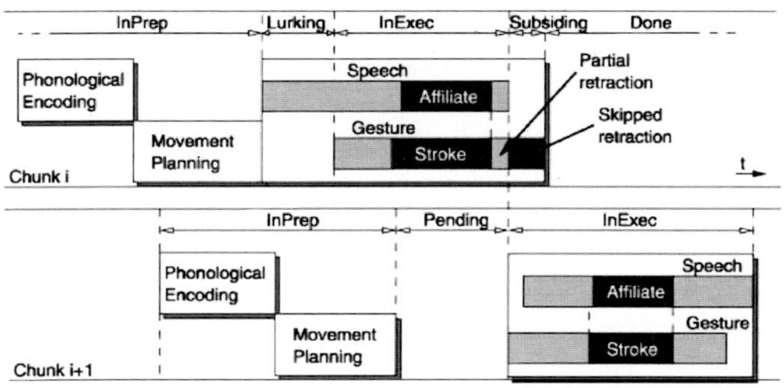

Fig. 2 Blackboards run through a sequence of processing states for incremental production of multi-modal chunks

This approach for gesture motor control is based on a hierarchical concept: During higher-level planning, the motor planner is provided with timed form features as described in the MURML specification. This information is then transferred to independent motor control modules. Such a functional-anatomical decomposition of motor control aims at breaking down the complex control problem into solvable sub-problems [18]. ACE [10] provides specific motor planning modules, amongst

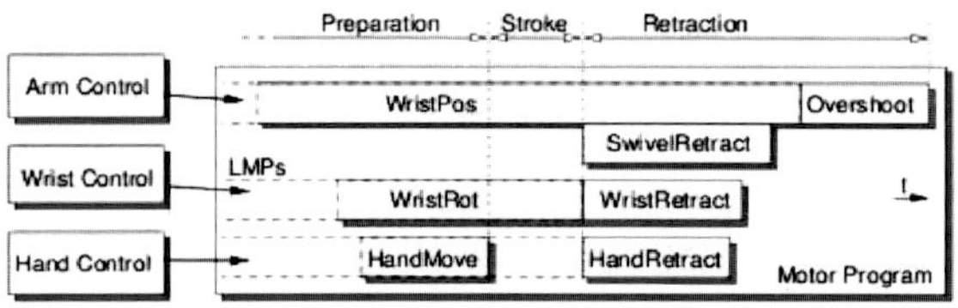

Fig. 3 Composition of motion control in Local Motor Programs (LMPs) for a hand-arm gesture

others, for the arms, the wrists, and the hands, which instantiate local motor programs (LMPs). These are used to animate required sub-movements and operate within a limited set of DOFs and over a designated period of time (Fig. 3). For each limb's motion, an abstract motor control program (MCP) coordinates and synchronizes the concurrently running LMPs for an overall solution to the control problem. The overall control framework, however, does not attend to how such sub-movements are controlled. To allow for an effective interplay of the LMPs within a MCP, the planning modules arrange them into a controller network which defines their potential interdependencies for mutual (de-)activation. LMPs are able to transfer activation between themselves and their predecessors or successors to ensure context-dependent gesture transitions. Consequently, they can activate or deactivate themselves at run-time based on feedback information on current movement conditions. Once activated, LMPs are continuously applied to the kinematic skeleton in a feedforward manner.

The on-the-fly timing of gestures is accomplished by the ACE engine as follows: The gesture stroke phase (the expressive 'core' phase) is set to accompany the co-expressive phase in speech (the 'affiliate') as annotated in the MURML specification. The ACE scheduler retrieves timing information about the synthetic speech at the millisecond level and defines the gesture stroke to start and end accordingly. These temporal constraints are automatically propagated down to each single gesture component (e.g. how long the hand has to form a certain shape). The motor planner then creates the LMPs that meet both the temporal constraints and the form constraints. The second aspect of scheduling, namely, the decision to skip preparation or retraction phases, emerges automatically from the interplay of motor programs at run-time. Motor programs monitor the body's current movements and decide when to activate themselves and to take action in order to realize the planned gesture stroke as scheduled. A retraction phase is skipped when the motor program of the next gesture takes over the control of the effectors from the previous program. This online scheduling creates fluent and continuous multi-modal behavior. It is possible because of the interleaved production of successive chunks of multi-modal behavior in ACE and has been employed successfully in several virtual humans.

4 Control Architecture for Robot Gesture

By re-employing existing concepts from the domain of virtual conversational agents, our goal is to similarly enable the robot to flexibly produce speech and co-verbal gesture at run-time. This requires a robot control architecture that combines conceptual representation and planning with motor control primitives for speech and arm movements, thereby endowing ASIMO with 'conceptual motorics'.

Since gesture generation with ACE is based on external form features as given in the MURML description, arm movement trajectories are specified directly in task space. The calculated vector information is passed on to the robot motion control module which instantiates the actual robot movement. For this purpose, externally formulated local motor programs (for wrist position and preparation/stroke of wrist flexion and swivel movement) are invoked first. Subsequently, inverse kinematics (IK) of the arm is solved on the velocity level using the ASIMO whole body motion (WBM) controller framework [5]. WBM aims to control all DOF of the humanoid robot by given end-effector targets, providing a flexible method of controlling upper body movement by specifying only relevant task dimensions selectively in real-time. For this purpose, task-specific command elements can be assigned to the command vector at any time, enabling the system to control one or multiple effectors while generating a smooth and natural movement. Redundancies are optimized regarding joint limit avoidance and self-collision avoidance. For more details on WBM control for ASIMO see [5].

Once IK has been solved for the internal body model provided for WBM control, the joint space description of the designated trajectory is applied to the real ASIMO robot. Due to constraints imposed by the robot's physical architecture and motor control, however, the inner states represented within the WBM controller might deviate from the actual motor states of the real robot during run-time. For this reason, a bi-directional interface for both efferent and afferent signaling is required. This is realized by a feedback loop, updating the internal model of ASIMO in WBM as well as the kinematic body model coupled to ACE at a sample rate r. Note, however, that for successful integration, this process needs to synchronize the two competing sample rates of the ACE framework on the one hand, and the WBM software controlling ASIMO on the other hand. Fig. 4 illustrates our proposed control architecture for robot gesture.

A main advantage of this approach to robot control in combination with ACE is the formulation of the trajectory in terms of effector targets in task space, which are then used to derive a joint space description using the standard WBM controller for ASIMO. An alternative method would aim at the extraction of joint angle values from ACE and a subsequent mapping onto the robot body model. This, however, might lead to joint states that are not feasible on the robot, since ACE was originally designed for a virtual agent application and does not entirely account for certain physical restrictions such as collision avoidance. As a result, solving IK using ASIMO's internally implemented WBM controller ensures safer postures for the robot. However, due to deviations from original postures and respective joint angles, the outer form of a gesture might be distorted such that its original meaning is

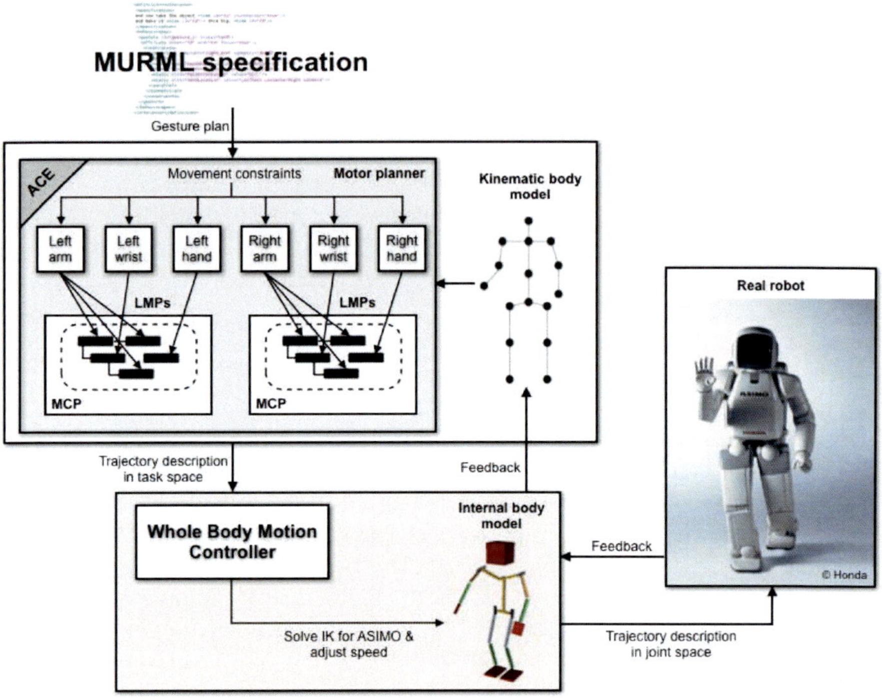

Fig. 4 Control architecture for robot gesture

altered. Therefore, whether and how the form and meaning of gestures are affected will be subject to further evaluation as work progresses.

5 Conclusion and Future Work

We presented a robot control architecture to enable the humanoid robot ASIMO to flexibly produce and synchronize speech and co-verbal gestures at run-time. The framework is based on a speech and gesture production model originally developed for a virtual human agent. Being one of the most sophisticated multi-modal schedulers, the Articulated Communicator Engine (ACE) allows for an on-the-spot production of flexibly planned behavior representations. By re-employing ACE as an underlying action generation architecture, we draw upon a tight coupling of ASIMO's perceptuo-motor system with multi-modal scheduling. This has been realized in a bi-directional robot control architecture which uses both efferent actuator control signals and afferent sensory feedback. Our framework combines conceptual, XML-based representation and planning with motor control primitives for speech and arm movements. This way, pre-defined canned behaviors can be replaced by conceptual motorics generated at run-time.

The requirement to meet strict synchrony constraints to ensure temporal and semantic coherence of communicative behavior presents a main challenge to our framework. Clearly, the generation of finely synchronized multi-modal utterances proves to be more demanding when realized on a robot with a physically constrained body than for an animated virtual agent, especially when communicative signals are to be produced at run-time. Currently, synchrony is mainly achieved by gesture adaptation to structure and timing of speech, obtaining absolute time information at phoneme level. To tackle this challenge the cross-modal adaptation mechanisms applied in ACE will be extended to allow for a finer mutual adaptation between robot gesture and speech.

Acknowledgements. The research project "Conceptual Motorics" is based at the Research Institute for Cognition and Robotics, Bielefeld University, Germany. It is supported by the Honda Research Institute Europe.

References

1. Bennewitz, M., Faber, F., Joho, D., Behnke, S.: Fritz – A humanoid communication robot. In: RO-MAN 2007: Proc. of the 16th IEEE International Symposium on Robot and Human Interactive Communication (2007)
2. Cassell, J., Bickmore, T., Campbell, L., Vilhjálmsson, H., Yan, H.: Human conversation as a system framework: desigining embodied conversational agents. In: Embodied Conversational Agents, pp. 29–63. MIT Press, Cambridge (2000)
3. Cassell, J., Vilhjalmsson, H., Bickmore, T.: Beat: the behavior expression animation toolkit. In: Proceedings of SIGGRAPH 2001 (2001)
4. Gibet, S., Lebourque, T., Marteau, P.F.: High-level specification and animation of communicative gestures. Journal of Visual Languages and Computing 12(6), 657–687 (2001)
5. Gienger, M., Janßen, H., Goerick, S.: Task-oriented whole body motion for humanoid robots. In: Proceedings of the IEEE-RAS International Conference on Humanoid Robots, Tsukuba, Japan (2005) (accepted)
6. Gorostiza, J., Barber, R., Khamis, A., Malfaz, M., Pacheco, R., Rivas, R., Corrales, A., Delgado, E., Salichs, M.: Multimodal human-robot interaction framework for a personal robot. In: ROMAN 2006: Proc. of the 15th IEEE International Symposium on Robot and Human Interactive Communication (2006)
7. Kopp, S., Bergmann, K., Wachsmuth, I.: Multimodal communication from multimodal thinking - towards an integrated model of speech and gesture production. Semantic Computing 2(1), 115–136 (2008)
8. Kopp, S., Wachsmuth, I.: A Knowledge-based Approach for Lifelike Gesture Animation. In: Horn, W. (ed.) ECAI 2000 - Proceedings of the 14th European Conference on Artificial Intelligence, pp. 663–667. IOS Press, Amsterdam (2000)
9. Kopp, S., Wachsmuth, I.: Model-based Animation of Coverbal Gesture. In: Proceedings of Computer Animation 2002, pp. 252–257. IEEE Pres, Los Alamitos (2002)

10. Kopp, S., Wachsmuth, I.: Synthesizing multimodal utterances for conversational agents. Computer Animation and Virtual Worlds 15(1), 39–52 (2004)
11. Kranstedt, A., Kopp, S., Wachsmuth, I.: MURML: A Multimodal Utterance Representation Markup Language for Conversational Agents. In: Proceedings of the AAMAS 2002 Workshop on Embodied Conversational Agents - let's specify and evaluate them, Bologna, Italy (2002)
12. Macdorman, K., Ishiguro, H.: The uncanny advantage of using androids in cognitive and social science research. Interaction Studies 7(3), 297–337 (2006)
13. McNeill, D.: Hand and Mind: What Gestures Reveal about Thought. University of Chicago Press, Chicago (1992)
14. Minato, T., Shimada, M., Ishiguro, H., Itakura, S.: Development of an android robot for studying human-robot interaction. In: Orchard, B., Yang, C., Ali, M. (eds.) IEA/AIE 2004. LNCS (LNAI), vol. 3029, pp. 424–434. Springer, Heidelberg (2004), `http://www.springerlink.com/content/rcvkmh0ucra0gkjb`
15. Reiter, E., Dale, R.: Building Natural Language Generation Systems. Cambridge Univ. Press, Cambridge (2000)
16. Sidner, C., Lee, C., Lesh, N.: The role of dialog in human robot interaction. In: International Workshop on Language Understanding and Agents for Real World Interaction (2003)
17. Wachsmuth, I., Kopp, S.: Lifelike Gesture Synthesis and Timing for Conversational Agents. In: Wachsmuth, I., Sowa, T. (eds.) GW 2001. LNCS (LNAI), vol. 2298, pp. 120–133. Springer, Heidelberg (2002)
18. Zeltzer, D.: Motor control techniques for figure animation. IEEE Computer Graphics Applications 2(9), 53–59 (1982)

Virtual Partner for a Haptic Interaction Task

Jens Hölldampf, Angelika Peer, and Martin Buss

Abstract. Interaction between humans and robots becomes more important in everyday life. In this work, a system is presented that creates a natural haptic human-robot interaction. A one degree of freedom dancing task is considered. The underlying model, which is based on a corresponding human-human interaction, replaces the male partner. The trajectories of the dancing steps are synthesized using a nonlinear system dynamics. A virtual fixture recreates the behavior of the interaction with a female partner. The implemented model is evaluated for different parameters based on objective measures.

1 Introduction

Today, robots are not restricted any more to industrial settings only, but are supposed to interact and assist humans in their everyday life. Achieving human-like behavior of the robot is a very difficult and complex task since small variations are immediately recognized as unnatural. This holds especially if physical contact with a human is required, like in rehabilitation, training, simulation of sports or physical guidance. Physical interaction requires a mutual information exchange of the two interacting partners that is by far not understood yet [6].

This paper focuses on the physical (haptic) interaction of two humans: In contrast to other studies, we do not examine a discrete or pointing task, but we consider dancing as a cyclic task that requires direct physical interaction. Dancing is characterized by an unbalanced distribution between the partners as the male is dominating the female, dancing steps are predefined and disturbances have to be balanced.

Jens Hölldampf · Angelika Peer · Martin Buss
Institute of Automatic Control Engineering, Technische Universität München, Munich, Germany
e-mail: {jens.hoelldampf,angelika.peer,mb}@tum.de

H. Ritter et al. (Eds.): Human Centered Robot Systems, COSMOS 6, pp. 183–191.
springerlink.com

The aim of this work is to create a virtual male dancing partner which imitates the behavior of a real human partner. We consider a virtual male partner to be more challenging than a female one. A virtual female partner requires the ability to understand the intentions of the partner in order to follow. For a virtual male partner, these intentions additionally have to be synthesized in a proper way. In order to achieve this, we first investigate human-human interaction. To simplify the analysis, an abstracted dancing scenario with one degree of freedom is considered. Then, in a second step, we propose a synthesis method for a virtual male partner. Finally, we compare real human-human interaction with human-robot interaction by means of pre-defined objective measures.

2 State-of-the-Art

Several studies concerning the analysis of human behavior while dancing are known from literature. Brown et al. examined the sensorimotor coordination while dancers were performing small-scale tango steps [1]. Gentry et al. investigated the haptic coordination between dancers with PHANToMs [3]. Takeda et al. built a female dancing robot which follows the male and tries to estimate the dance steps based on Hidden Markov Models [8]. In a discrete task, Reed et al. looked at the physical interaction between two humans [6] and showed that simply replaying the behavior of one recorded human is not suitable for the imitation of a real human-human interaction [5].

To our best knowledge, no synthesis of a haptic enabled male dancer and its implementation as an interactive robot has been shown before. This is considered a very challenging task because of the high variability of human motion and force patterns. When e.g. capturing motion data, it is very unlikely that the same trajectory will be recorded twice. Although there can be variations with respect to position, amplitude and time, a human will still recognize the movement as normal and natural as long as certain characteristics are maintained. Thus, the main challenge addressed in this paper is to model typical dancing steps in real human dancing and implementing them in an interactive robotic partner.

3 Haptic Interaction Task

In human-human interaction, the two humans perform the dancing task together. The recorded data are used to synthesize the male partner. In order to standardize human-human as well as human-robot interaction, the interaction is realized by means of two virtually coupled haptic interfaces, one for the female and one for the male. In human-robot interaction, which is the aim of the synthesis, the female uses one haptic interface to interact with a virtually rendered male partner, see Fig. 1.

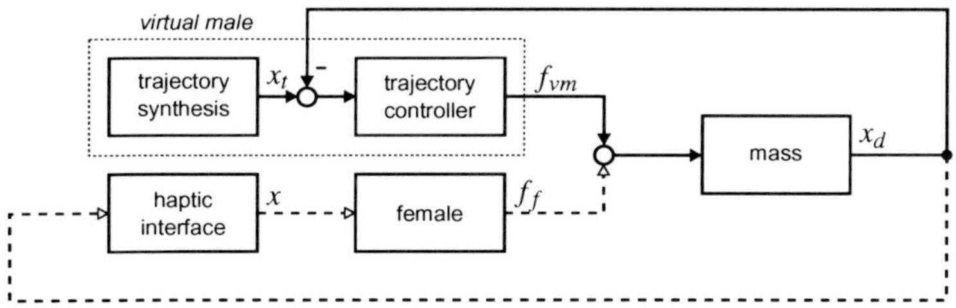

Fig. 1 Block diagram of virtual male interaction with female

As a real dancing scenario is very complex and thus difficult to analyze we first abstracted the task to a one degree of freedom dancing task. Three dancing steps, *left*, *right*, and *full*, are performed as rhythmic movements corresponding to music provided via headphones. The beat of the music serves as an external synchronization between the partners.

Two direct drive linear axes, Thrusttube module of Copley Controls Corp., are used as haptic interfaces. The linear axes are equipped with one DOF force sensors to measure the individual forces applied by the humans. A standard Linux PC using the Real Time Application Interface (RTAI) and the Real-Time Workshop of MATLAB/Simulink controls the haptic interfaces via a multichannel analog and digital I/O card (Sensoray 626).

To simulate a real dancing situation, instructions of the current dancing step were given only to the male partner: A bar displayed on his screen indicated the next dancing step, whereby the transitions between two steps were announced in advance. The whole setup is shown in Fig. 2.

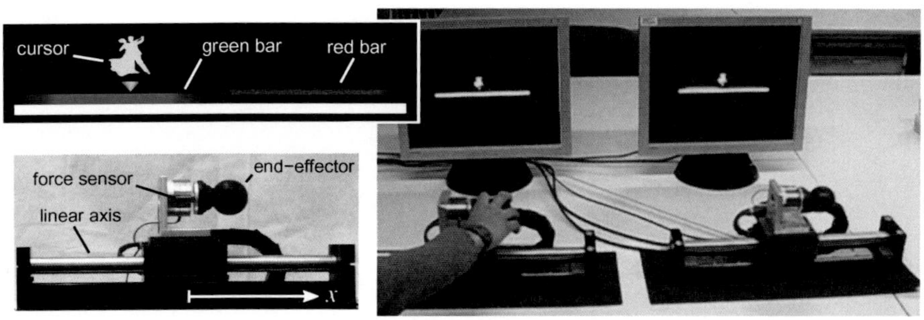

Fig. 2 Setup for human-human interaction with two screens and haptic interfaces, female grasping one end-effector. Screen shot with the green bar indicating the current dancing step in the left half and the red bar announcing the next transition to the right half. Haptic interface consisting of linear axis, force sensor, and end-effector.

4 Analysis of Human-Human Interaction

To obtain data of the interaction between two humans and to synthesize the male partner, couples of male and female were recorded while they performed the already described dancing task. Fig. 3 shows such a typical recording for the three dancing steps in the $x/\dot{x}$ plane. The trajectories are similar for each step, but a certain variation has to be considered when synthesizing a virtual male partner.

Dancing does not only consist of trajectories, it also requires haptic communication between the dancing partners. To analyze this communication, we consider the interactive force applied during interaction. The interactive force f_i is defined as the force which does not contribute to the movement

$$f_i = \begin{cases} 0 & sgn(f_m) = sgn(f_f) \\ f_m & sgn(f_m) \neq sgn(f_f) \wedge |f_m| \leq |f_f| \\ -f_f & sgn(f_m) \neq sgn(f_f) \wedge |f_m| > |f_f| \end{cases} \quad (1)$$

where f_m denotes the force applied by the male and f_f the one applied by the female [2]. We define the sign of the interactive force to be oriented in the same direction as the male force. The interactive force applied during dancing is shown in Fig. 4. We consider the interactive force to be correlated with the amount of information to be exchanged between partners, whereby low values indicate that the two partners can interpret the intentions of their partner and thus smoothly interact with each other. The interaction with a natural virtual male partner should thus lead to similar interactive forces as found in real human-human interaction.

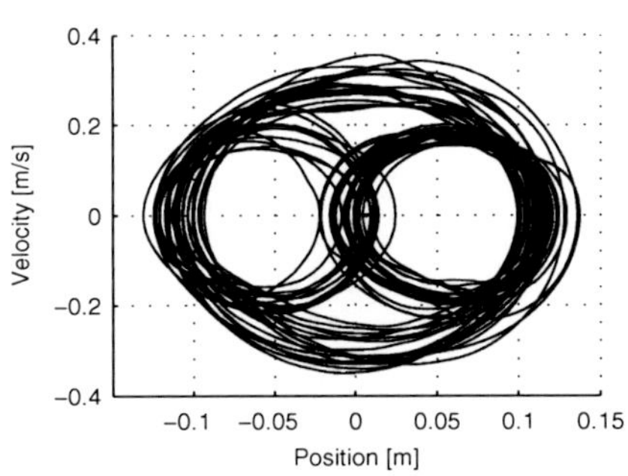

Fig. 3 Recorded trajectory with the three circles corresponding to the dancing steps *left*, *right*, and *full*

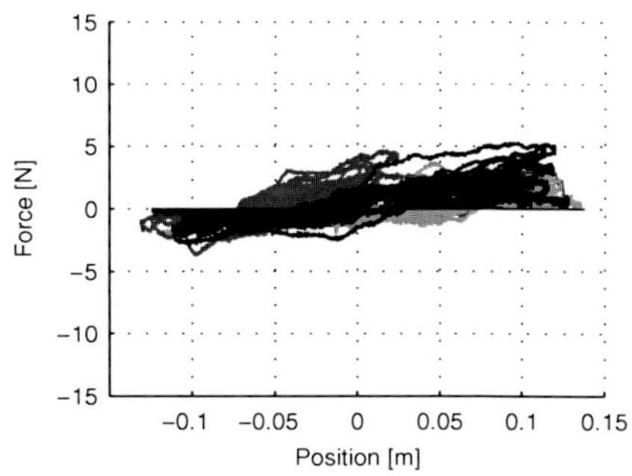

Fig. 4 Recorded interactive force for the three dancing steps ***left***, *right*, and ***full***

5 Synthesis of Virtual Partner

Dancing requires following a desired trajectory and correcting position deviations by means of a trajectory controller, see Fig. 1. This can be compared to the concept of virtual fixtures [7] which introduces geometric constraints in order to guide the human through a task. Implementing a male robotic partner in form of a virtual fixture requires i) defining (synthesizing) a desired trajectory and ii) modulating the strength of the guidance depending on the actual task requirements. In this paper we only address the first topic while assuming constant strength of the trajectory controller.

5.1 Trajectory Synthesis

Dancing steps can be described as trajectories. They are represented as a state vector $x(t) \in \mathbb{R}^N$ in continuous or $x[k] \in \mathbb{R}^N$ in time-discrete systems with N states. The simplest, but not very efficient way of defining a desired trajectory is to record position trajectories in real human-human interaction and store all single data points. We consider a more efficient way of storage by adopting the approach of Okada et al. [4]. They proposed a method to generate a nonlinear system dynamics based on a trajectory. We apply this principle to synthesize and store the recorded position trajectories of an interaction between two humans. The stored trajectory represents the mean characteristics of the recorded ones. In contrast to thousands of stored data points for a typical dancing scenario we end up with a few hundred parameters to be stored for all dancing steps, which results in a tremendous reduction of memory requirements. The principle of the applied method is outlined in the following paragraphs.

We consider the desired trajectory C which consists of M samples c. These samples are measured positions over time.

$$C = \left[\, c[1] \quad c[2] \; \cdots \; c[M] \,\right] \tag{2}$$

$$c[k] = \left[\, c_1[k] \quad c_2[k] \; \cdots \; c_N[k] \,\right]^T \tag{3}$$

The dynamic system has the general form

$$x[k+1] = x[k] + f(x[k]) \tag{4}$$

where f denotes the system dynamics. As the system in (4) is time-discrete, f generates an implicit time dependency. We assume a constant sampling time.

To obtain the system dynamics, a vector field is constructed around the curve. Fig. 5 illustrates this principle. The vectors are pointing towards the curve to assure that the system state is attracted by the desired trajectory.

Around each measured sample $c[k]$ of the curve, several points η_i can be defined (Fig. 6). $\delta[k]$ and $\gamma[k]$ are perpendicular to $\Delta c[k]$.

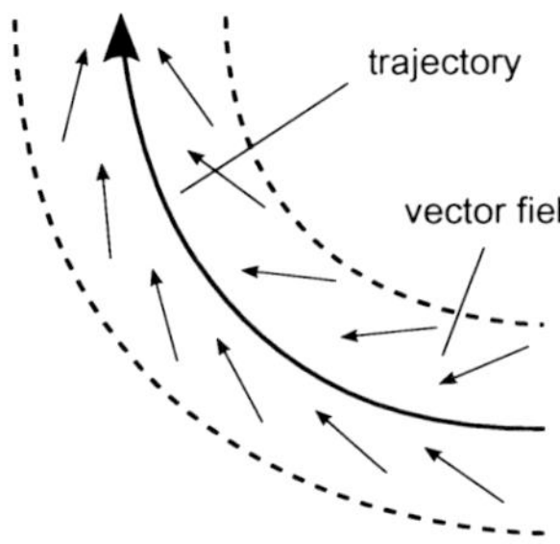

Fig. 5 Trajectory with attracting vector field

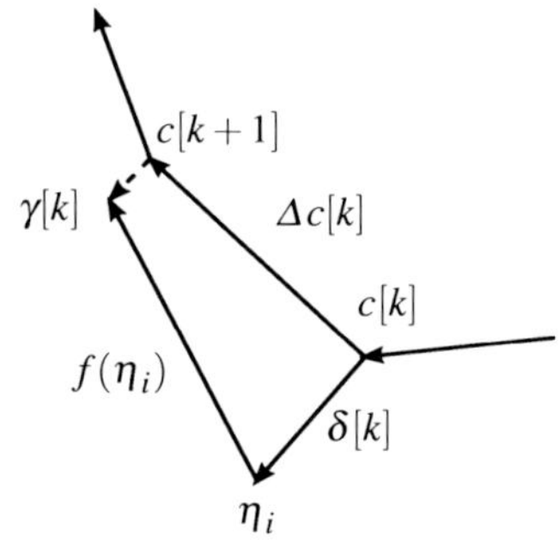

Fig. 6 Curve with attracting vector field

In order to determine the system dynamics $f(x[k])$, the vector field $f(\eta_i)$ is approximated by a polynomial expression

$$f(\eta_i) = \boldsymbol{\Phi}\theta(x) \tag{5}$$
$$\theta(x) = [\, x_1^l \quad x_1^{l-1}x_2 \quad x_1^{l-2}x_2^2 \cdots 1\,]^T \tag{6}$$

The variable l denotes the degree of the polynomial and $\boldsymbol{\Phi}$ is a constant parameter matrix. It is determined by applying the least square method

$$\boldsymbol{\Phi} = F\Theta^{\#} \tag{7}$$
$$F = [\, f(\eta_1) \quad f(\eta_2) \cdots f(\eta_n)\,] \tag{8}$$
$$\Theta = [\, \theta(\eta_1) \quad \theta(\eta_2) \cdots \theta(\eta_n)\,] \tag{9}$$

Okada et al. also proposed an extension which incorporates an input signal into the system [4]. The input signal allows to influence the overall system behavior by switching between different dynamic subsystems. It is consequently possible to segment the trajectory into different states, e.g. dancing steps, by adopting different levels of the input signal and synthesize it again by applying the appropriate signal levels. For this purpose, the overall trajectory has to be approximated with different vector fields for each state, respectively input signal.

The dynamic system (4) is modified as follows:

$$x[k+1] = x[k] + g(u[k], x[k]) \tag{10}$$

whereby the vector $u[k]$ contains the input signal. It is appended to C in (2) and approximated in the same way as $f(x[k])$. This principle provides a convenient way to change between dancing steps as the steps can be faded into each other at the switching points.

5.2 Simulation

Fig. 7 shows the three dancing steps generated by the dynamic system at a sampling rate of 100Hz. The vector field with its attracting behavior for the dancing step *right* is shown in Fig. 8.

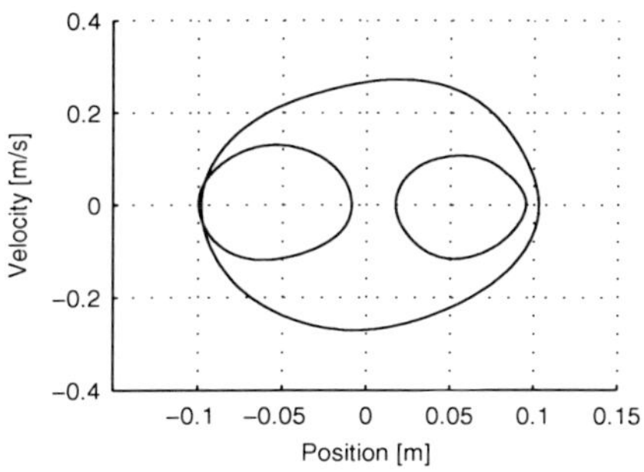

Fig. 7 Synthesized steps *left*, *right*, and *full*

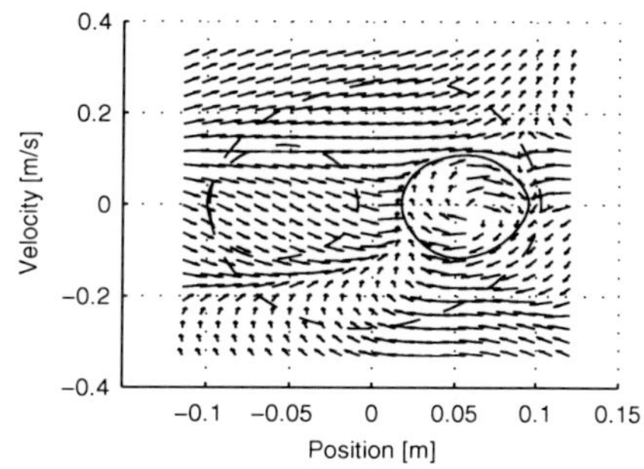

Fig. 8 Step *right* with corresponding vector field

6 Evaluation of Virtual Male Partner Model

This section aims at evaluating the synthesized male partner. For this purpose the data acquired during human-human interaction is compared with the data acquired during human-robot interaction. Different gains of the implemented trajectory controller were tested. Figs. 9 and 10 show the data of the achieved interaction between a female and a virtual male partner assuming a high-gain trajectory controller.

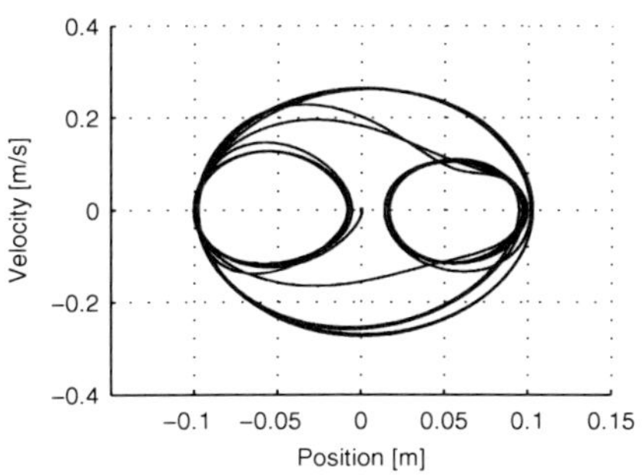

Fig. 9 Trajectory with high-gain controller

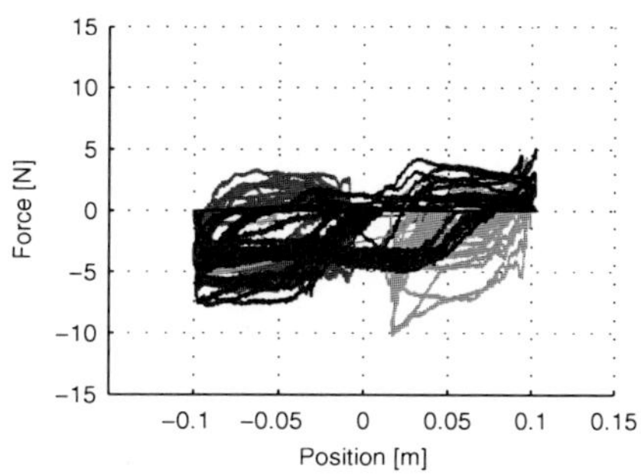

Fig. 10 Interactive force with high-gain controller

Compared to the real human-human interaction trials presented in section 4, the variations of the trajectory in the $x/\dot{x}$ plane are smaller and the amount of interactive force is higher.

On the other hand, the trajectory and the interactive force of a female interacting with low-gain trajectory controller are shown in Figs. 11 and 12.

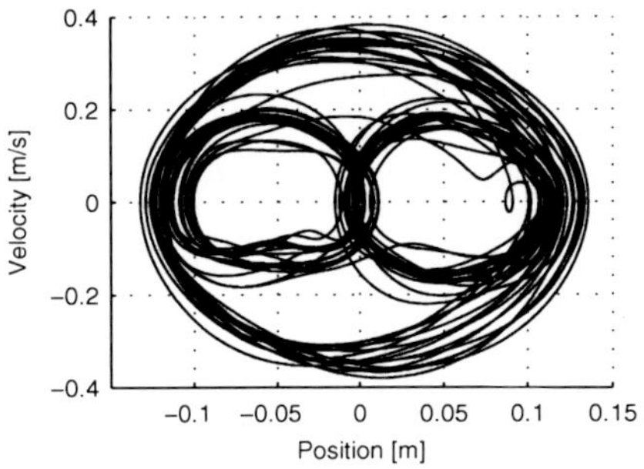

Fig. 11 Trajectory with low-gain controller

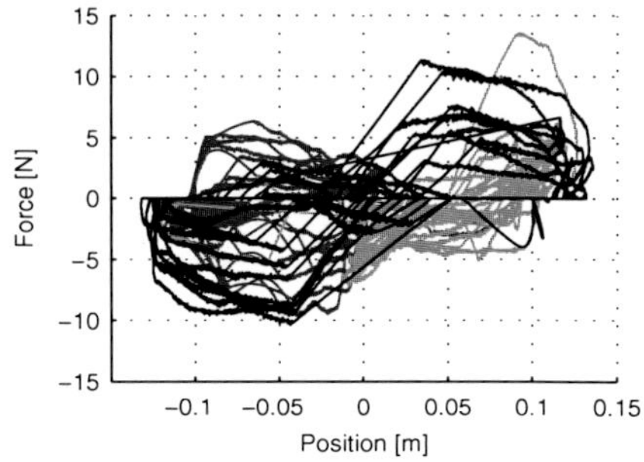

Fig. 12 Interactive force with low-gain controller

In this case, the trajectories in the $x/\dot{x}$ plane are similar to the human-human interaction trials but the interactive force is again higher. We interpret the bigger amount of interactive forces in both cases as a lack in the understanding of the model's behavior as it does not respect the subtle haptic signals found in human-human interaction.

7 Conclusions and Future Work

This paper presents an approach to create a natural human-robot interaction based on the knowledge gained from the corresponding human-human interaction. A one degree of freedom dancing task is considered. Based on the recording of two humans, a model is built to recreate the behavior. Different control parameters of the implemented trajectory controller are tested to validate the model. The results show that the synthesized trajectories can be similar to the human-human interaction. However, the interactive forces are higher during the human-robot interaction. This increase indicates a lack of communication between female and the virtual male partner. Both partners seem to act against each other instead of coordinating their common movement. Hence, we conclude that a constant virtual fixture is insufficient to create a natural human-robot interaction in the selected dancing scenario. Thus, the virtual partner needs the ability to adapt his behavior to the behavior of the female. In the next step, a trajectory controller with variable parameters will be investigated. Beside comparison on an objective level, a Turing test will be conducted to further validate the human-robot interaction by means of subjective measures.

Acknowledgements. We gratefully acknowledge the help of Raphaela Groten in the design of the experimental paradigm for the analysis of the human-human interaction in dancing. This work is supported in part by the ImmerSence project within the 6th Framework Programme of the European Union, FET - Presence Initiative, contract number IST-2006-027141. For the content of this paper the authors are solely responsible for, it does not necessarily represent the opinion of the European Community, see also www.immersence.info.

References

1. Brown, S., Martinez, M.J., Parsons, L.M.: The neural basis of human dance. Cerebral Cortex 16(8), 1157–1167 (2006)
2. Feth, D., Groten, R., Peer, A., Buss, M.: Control-theoretic model of haptic human-human interaction in a pursuit tracking task. In: Proceedings of the 18th IEEE International Symposium on Robot and Human Interactive Communication, ROMAN (accepted, 2009)
3. Gentry, S., Murray-Smith, R.: Haptic dancing: Human performance at haptic decoding with a vocabulary. In: Proc. IEEE International Conference on Systems, Man and Cybernetics, vol. 4, pp. 3432–3437 (2003)
4. Okada, M., Tatani, K., Nakamura, Y.: Polynomial design of the nonlinear dynamics for the brain-like information processing of whole body motion. In: Proc. IEEE International Conference on Robotics and Automation, ICRA 2002, vol. 2, pp. 1410–1415 (2002)
5. Reed, K.B., Patton, J., Peshkin, M.: Replicating human-human physical interaction. In: Proc. IEEE International Conference on Robotics and Automation, pp. 3615–3620 (2007)
6. Reed, K.B., Peshkin, M., Hartmann, M.J., Patton, J., Vishton, P.M., Grabowecky, M.: Haptic cooperation between people, and between people and machines. In: Proc. IEEE/RSJ International Conference on Intelligent Robots and Systems, pp. 2109–2114 (2006)
7. Rosenberg, L.B.: Virtual fixtures: Perceptual tools for telerobotic manipulation. In: Proc. IEEE Virtual Reality Annual International Symposium, pp. 76–82 (1993)
8. Takeda, T., Hirata, Y., Kosuge, K.: Dance step estimation method based on hmm for dance partner robot. IEEE Transactions on Industrial Electronics 54(2), 699–706 (2007)

Social Motorics – Towards an Embodied Basis of Social Human-Robot Interaction

Amir Sadeghipour, Ramin Yaghoubzadeh, Andreas Rüter, and Stefan Kopp

Abstract. In this paper we present a biologically-inspired model for social behavior recognition and generation. Based on an unified sensorimotor representation, it integrates hierarchical motor knowledge structures, probabilistic forward models for predicting observations, and inverse models for motor learning. With a focus on hand gestures, results of initial evaluations against real-world data are presented.

1 Introduction

For human-centered robots to be able to engage in social interactions with their users, they need to master a number of daunting tasks. This includes, e.g., robust and fast recognition and understanding of interactive user behavior, human-acceptable expressivity, joint attention, or incremental dialog with mutual adaptivity. In humans such capabilities are assumed to rest upon an embodied basis of social interaction – direct interactions between perception and generation processes (*perception-behavior expressway* [6]) that support mirroring or resonance mechanisms [15] to process social behavior at different levels, from kinematic features to motor commands to intentions or goals [8]. Research in social robotics has increasingly started to adopt such principles in its work on architectures and interaction models (e.g. [3, 5]). Against this background we present our work towards "social motorics", modeling a resonant sensorimotor basis for observing and using social behavior in human-robot interaction. With a focus on hand-arm gestures, we describe a probabilistic model that exploits hierarchical motor structures with forward and inverse models in order to allow resonance-based processing of social behavior. We start in Section 2 with a review of related work on gesture learning and recognition. In Section 3 we introduce our overall computational model and detail the

Amir Sadeghipour · Ramin Yaghoubzadeh · Andreas Rüter · Stefan Kopp
Sociable Agents Group, CITEC, Bielefeld University
e-mail: {asadeghi, ryaghoub, arueter, skopp}@techfak.uni-bielefeld.de

H. Ritter et al. (Eds.): Human Centered Robot Systems, COSMOS 6, pp. 193–203.
springerlink.com

employed forward and inverse models in Section 4 and 5, respectively. The probabilistic modeling of interactions between perception and generation is described in Section 6 and, afterwards, we present in Section 7 results of how the model performs at simulating resonances during perception of gestural behavior. Finally, Section 8 gives a conclusion and summery of future works.

2 Related Works

The growing interest in developing social artificial agents requires abilities for perception, recognition and generation of gestures as one of the non-verbal interaction modalities. The recognition process is concerned with the analysis of spatio-temporal features of the hand movements and is mainly treated as pattern classification with subsequent attribution of meaning. Many studies apply probabilistic approaches to classify hand gesture trajectories. Hidden Markov models have been widely applied as an efficient probabilistic approach to work with sequence of data [1, 4, 7]. However, the focus of those approaches is on pattern recognition separated from the attribution of meaning, and they rely on hidden variables which do not directly correspond to the agent's own action repertoire. For recognizing transitive actions, in which the goal of an observed action is often visually inferable, hierarchical models are used to analyze the perceived stimuli in a bottom-up manner towards more abstract features and, consequently, goals of those actions [2, 11, 17].

Furthermore, imitation mechanisms (overt or covert ones) are widely used for learning and reproducing behaviors in artificial agents. For instance, the MOSAIC model applies forward and inverse models to predict and control movements in a modular manner [9]. Others [10, 17] have worked on hierarchical MOSAIC models towards more abstract levels of actions. However, none of these models has adopted imitation mechanisms to attain perception-action links using a shared motor representation – a hallway of what we assume here to be the basis of social motorics.

3 Resonant Sensorimotor Basis

In our work we aim at modeling perceptuo-motor processes that enable a robot, on the one hand, to concurrently perceive, recognize and *understand* motor acts of hand-arm gestures and learn them by imitation (cf. [12, 14]). That is, the model is to process the robot's perceptions automatically, incrementally, and hierarchically from hand and arm movement observation toward understanding and semantics of a gesture. In result, the robot's motor structures are to start to "resonante" to the observation of corresponding actions of another structurally congruent agent (either human or artificial). On the other hand, the model is designed to allow the robot to *generate* gestures in social interaction from the same motor representation.

Overall, the model connects four different structures (Fig. 1): preprocessing, motor knowledge, forward models, and inverse models. We presume that some kind of perceptual processing has identified a human's body parts as relevant for hand-arm gesture. Now, the preprocessing module receives continuous stimuli about the user's hand postures (finger configurations) and wrist positions in the user's effector space. Since in our framework the received sensory data are already associated with corresponding body parts (left/right arm and left/right hand) of the human demonstrator (cf. Section 7), and since we assume the robot to be anthropomorphic, body correspondence can be established straight-forward. A *body mapping* submodule maps the perceived data from the human-centered coordinate system into the robot's body frame of reference. The *sensory memory* receives the transformed visual stimuli at each time step and buffers them in chronological order in a *working memory*.

The motor knowledge structures encode the robot's competence to perform certain gestural behaviors itself – more specifically, to perform the required movements of the relevant body parts. This knowledge is organized hierarchically. The lowest level contains *motor commands* (MC) required for the single movement segments (cf. [13]). Data for each of the four relevant body parts is stored in directed graphs, the nodes of which are intermediate states within a gestural movement; the edges represent the motor commands that lead from one state to another. The next level, also present for each body part, consists of *motor programs* (MP) that cluster several sequential MCs together and represent paths in the motor command graph. Each MP stands for a meaningful movement, i.e., a gestural performance executed with the respective body part. Since gestures typically employ both the hand as well as the more proximal joints (elbow, shoulder), often even in both arms, all contributing body parts need to be controlled simultaneously. Furthermore, gestures are generally not restricted to a specific performance but have some variable features which, when varied, do not change the meaning of the gesture but merely the way of performing it. Thus, a social robot must be able to cluster numerous instances of a gestural movement into a so-called "schema", which demarcates the stable,

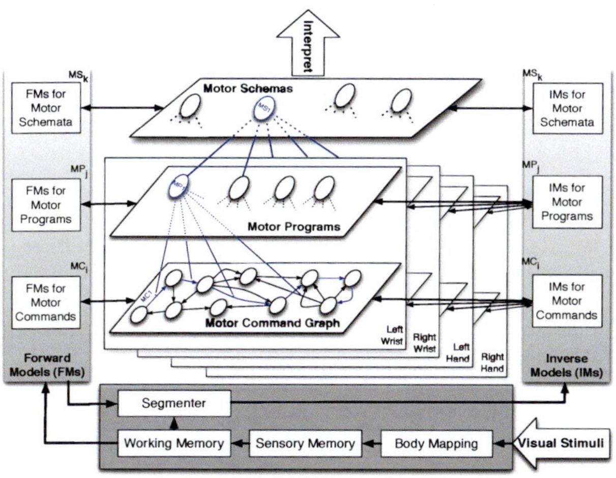

Fig. 1 Outline of the structure of a sensorimotor basis for social motorics

mandatory features from the variable features. For example, a waving gesture has a number of determinant features (hand lifted, palm facing away from the body, reciprocating motion in the frontal body plane), while bearing a number of variable features that mark its context-dependency or manner of execution (e.g., number of repetitions, speed, handedness, height). Therefore, we define *motor schemas* (MS) as a generalized representation that groups different, familiar performances in relevant body parts (MPs) of a gesture into a single cluster. Such a generalization process can foster the understanding and imitation of behavior in several ways. First, by combining different body parts into an MS, a gesture can be recognized more robustly combining information about different body parts. Second, the concept of motor schemas elevates the problem of interpreting a gesture from the complex motor level to a more abstract, yet less complex level, namely schema interpretation. Third, a robot can retain its own personal form of performing a gesture while being able to relate other performances of the same gesture to the same schema.

The third structure is formed by forward models. Such models are derived from the robot's motor knowledge at each level. While observing a behavior, they run *internal* simulations in order to predict how the behavior would continue for each possible explanation considered. By evaluating this prediction against the actual percepts at each time step, this structure is able to determine how well individual motor commands, programs or schemas correspond to the observed behavior. If there is no sufficiently corresponding representation, the processing switches to the final structure, the inverse models. These are responsible for learning, i.e. analyzing the movements of a behavior and augmenting the robot's motor knowledge at all three levels correspondingly. To this end, a *segmenter* on the lowest level decomposes received movements into (nearly) planar segments based on their kinematic features (i.e. velocity profile and direction changes).

In the following, we will present a probabilistic approach to model these three core structures (motor knowledge, forward models and inverse models) for intransitive, gestural movements.

4 Forward Models

The classification of observed input into levels of increasing abstraction, as described above, is achieved by matching it with simulations performed according to the receiver's own motor repertoire. This simulation and matching is performed by the *forward models* in a probabilistic way. The aim is to find a set of hypotheses from the repertoire that can explain the observed input. For each hypothesis considered, a class of functors termed *predictors* constructs a probability density function for the input likelihood for an arbitrary time in the future (the *prediction*), under the condition that the motor component associated with that hypothesis were to be the one producing the observations. The result of the ongoing evaluation of these expectations against the actual evidence is assumed to reflect automatic "resonances" in the robot's motor hierarchy. As demanded for an embodied basis of social interaction, this method therefore consequently carries the assumption of a correspondence between the motor repertoires of the human user and the social robot.

Based on the predictors, any observation is evaluated against the densities predicted for that time. This yields a measure of *explainability* of the data under the assumptions implied by the hypothesis (*diagnostic support*). The resulting *a-posterori* performances of the hypotheses are then compared using Bayes' theorem, taking into account the probability of certain motion primitives as provided in the form of prior distributions, which can be influenced by higher levels. The explicit posterior distributions are also used for a suitable pruning of the search space, allowing both the retention of plausible hypotheses while at the same time discarding those hypotheses deemed negligible. Full probabilistic forward models for the levels of motor commands, motor programs and motor schemas for wrist trajectories in 3D space have been implemented and tested (cf. Section 7). The forward model at the motor command level makes use of distribution functions formed by the convolution of a configurable Gaussian kernel along parts of the possible trajectories as spanned by consecutive motor commands. The covered 3D space is also a function of time, which is addressed by another configurable distribution function relating the individual tolerated speed variance to a path segment along the trajectory. This set of variable-density "tubular clouds" (Fig. 2(d)) is utilized as hypothesis-dependent likelihood functor $P(\mathbf{o}_t|c)$. Formula 1 details the generation of posteriors with a *prior feedback* approach (using the previous posterior as prior $P_{T-1}(c)$).

$$P_T(c|\mathbf{o}) := \frac{1}{T}\sum_{t=t_1}^{T} P(c|\mathbf{o}_t) = \frac{1}{T}\sum_{t=t_1}^{T} \alpha_c P_{T-1}(c)P(\mathbf{o}_t|c) \tag{1}$$

The forward models at higher levels work similarly in a Bayesian fashion and consider the observation and hypotheses from the lower levels. The motor program hypotheses (Formula 2) contain additionally the likelihood functor $P(c|p)$, which is modeled as a simple discrete Gaussian probability distribution along the according motor commands at each time step.

$$P_T(p|C,\mathbf{o}) := \frac{1}{T}\sum_{t=t_1}^{T} \alpha_p P_{T-1}(p) \sum_{c\in C} P(\mathbf{o}_t|c)P_t(c|p) \tag{2}$$

A motor schema clusters different performances of a gesture with certain variable features. The corresponding forward model at this level contains a new likelihood functor $P(p|s)$ that equals one, only if the according motor program, p, is clustered to the given motor schema, s. In addition, the forward model contains both likelihood models of the lower levels. Hence, the parameters of those likelihood functors (variances) can be set by each motor schema in order to take the variable performance features into consideration, i.e. velocity and position of motor commands or repetition of a movement segment. Furthermore, motor schemas determine which body parts contribute to the performance by applying an AND- or OR-relation (sum, product or a combination of both) to combine the prediction probabilities adequately. Formula 3 considers the case of performing a gesture using all four body parts: right/left arm (*rw*/*lw*) and right/left hand (*rh*/*lh*).

$$P_T(s|\mathbf{C},\mathbf{P},\mathbf{o}_{lw},\mathbf{o}_{rw},\mathbf{o}_{lf},\mathbf{o}_{rf}) := \frac{1}{T}\sum_{t=t_1}^{T}\alpha_s P_{T-1}(s) \prod_{i\in\{rw,lw,rh,lh\}} \sum_{p\in P} P(p_i|s) \sum_{c\in C} P(\mathbf{o}_{i,t}|c_i)P_t(c_i|p_i) \tag{3}$$

In future work, biological constraints and proprioceptive information will be applied during the simulation. This will allow a richer attribution to an observed movement (e.g. as being effortful and hence emphasized). Furthermore, the forward models will also be used to assess the feasibility of hypothetical motor commands for the robot before enriching the repertoire or executing them for (true) imitation.

5 Inverse Models

Whenever a novel behavior is observed, i.e. the forward models have failed to yield a sufficient explainability from the known repertoire, an *inverse model* takes over. It is in charge of formulating motor structures that can reconstruct the novel observation at the respective level of representation, thereby allowing for extension of the robot's repertoire. In our approach, the learning of gestures at the MC level uses a self-organizing feature map (SOM) to map observations over time onto a lower dimensional grid of neurons that represent prototypes, derived from gestures perceived in the past. These prototypes are used for classification and the generation of motor commands that form the repertoire of the robot (Fig. 2(a-c)). The best-matching neuron is determined via a "winner-takes-all" approach and its neighbourhood is adjusted by the difference between the input and the best match. The emergent map features smooth transitions between adjacent prototypes.

The inverse model for motor commands operates on movement segments. The input data are present as a sequence of nearly planar 3D segments, which are first projected into 2D using PCA, transformed into a common coordinate frame, and sampled in equidistant intervals. This normalization, which is inverted during final reconstruction, allows for the comparison of prototypes and input necessary for classification and training of the SOM. We use a randomly initialized, dynamical and online-learning SOM, which enables classification while in training. The dynamic learning process is controlled in order to prevent overfitting of the map, and eventually suspended until new input is presented.

From the winner neurons for the single movement segments, correspondingly parameterised MCs can be directly computed (cf. [13]) and imparted to the motor command graph. This will also insert a new MP, if there is none which consists of the sequence of the winner MCs. The motor schema level reaches into the context of gesture use and needs to cluster instances of gesture performance. This decision is subject of ongoing work, in which we consider how imitation with informative feedback from a human interlocutor can scaffold the learning of invariant features of a gesture schema and fitting of the likelihood parameters of the corresponding forward model at this level when new performances of familiar gestures are observed.

6 Resonance-Based Behavior Processing

The proposed model employs one and the same motor knowledge to guide the recognition of familiar hand-arm gestures, and as repository of motor commands and programs in the self-generation of behavior. Such a direct link between perception and action is assumed to underlie the evident cross-activation and influence of the two processes. The resulting mirroring of actions made by another individual is assumed to be fundamental to social understanding and embodied communication [15]. Such resonances in sensorimotor structures can enable many mutualities abundant in social interaction [6], e.g. non-conscious mimicry when leaking through to execution, or alignment when leaving traces that affect behavior production.

In our model, perception-induced resonances are the posterior probabilities of valid hypotheses. It is also simulated how such resonances percolate upwards, from single motor commands to higher-level structures, as well as how higher levels may affect and guide the perception process at lower levels over the next time steps. These processes are accounted for by computing the posteriors using Bayes' law and inserting prior probabilities for each motor component which depend on three criteria: (1) the number of candidate hypotheses (assigning the default priors); (2) the a-posteriori from upper levels (cf. Section 4); (3) the posterior probability in the previous time step, since we apply the prior-feedback method to model time dependency between sequential evidences. The combination of these priors affects the activation of the corresponding motor component during perception.

Except for the first one, these criteria also carry on information about the last perceived gesture. Therefore, these priors are not directly reset to their default values after perception, but decline following a sigmoidal descent towards the default a-priori. When the robot, as advocated here, uses the same motor knowledge and consequently the same prior probabilities while selecting proper motor components for producing its own behavior, the robot tends to favor those schemas, programs, and motor commands that have been perceived last. The other way around, the model also allows to simulate "perceptual resonance": choosing a motor component for generation increases its prior probability temporarily, biasing the robot's perception toward the self-generated behavior – another suggested mechanism of coordination in social interaction [16].

7 Results

The proposed model for resonance-based gesture perception has been implemented and tested with real-world gesture data in a setup with a 3D time-of-flight camera

(SwissRanger^{TM}SR4000[1]), which can be easily mounted to any mobile robot, and the marker-free tracking software iisu[2].

Inverse model. The performance of the applied SOM at the MC level depends on the training data and parameters. The result of a trained 4×4 SOM is shown in Fig. 2(a-c), after segmenting the observed wrist trajectory of a figure "3" drawn in the air.

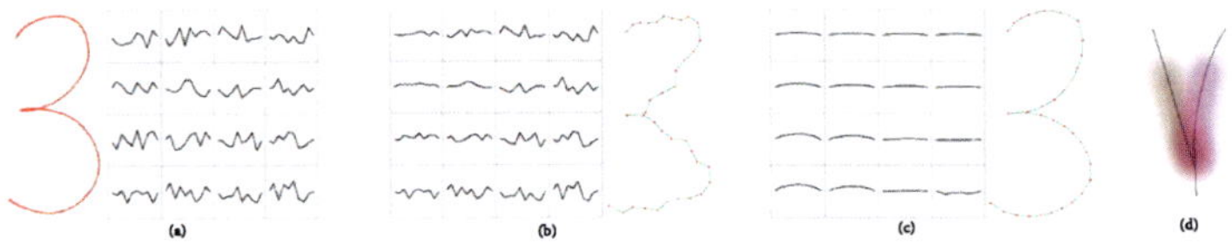

Fig. 2 (a) The performed figure "3" trajectory and a randomly initialized SOM; (b) the SOM and mapped trajectory after 10 training iterations and (c) after 150 iterations; (d) visualization of a time-dependent likelihood function $P(o_t|h)$ used by the forward models

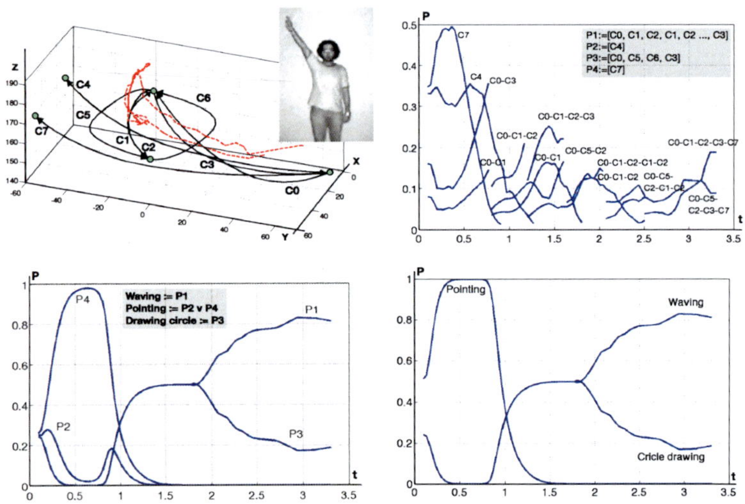

Fig. 3 Simulation results: *top-left:* motor command graph with observed trajectory overlaid (dashed line); *top-right/bottom-left/bottom-right:* changing probabilities of the hypotheses currently entertained on the three motor levels (*commands/programs/schemas*)

Forward models. In the following example, the motor knowledge (Fig. 3 top-left) was built based on observation of several performances of four different gestures: waving, drawing a circle, and two variants of pointing upwards. Fig. 3 shows how

1 http://www.mesa-imaging.ch

2 http://www.softkinetic.net

the confidences of alternative hypotheses at all three motor levels are evolving *during* the perception of another waving gesture.

At each time point in the observation, one hypothesis corresponds to the most expected movement component. Depending on the number of hypotheses, the maximum expectation value changes over time and the winner threshold is adopted respectively. As shown in Fig. 3 (bottom), the hypotheses first indicate that the observation is similar to a familiar pointing gesture (c_7). Therefore the robot thinks that the user is going to point upwards (p_4). However, after one second the user starts to turn his hand to the right. Thus, the expectation values of the motor commands c_1 and c_5 increase. Consequently, the gestures (p_1 and p_3) attain higher expectancies but the robot still cannot be sure whether the user is going to draw a circle or wave. After about two seconds the movement turns into swinging, which is significantly similar to the waving gestures (p_1) known to the robot. In result, the robot associates the whole movement with the waving schema and can now, e.g., execute a simultaneous imitation using his motor commands in the winning motor program.

8 Conclusion and Outlook

We presented our work towards the establishment of a sensorimotor foundation for social human-robot interaction, guided by neurobiological evidence regarding motor resonance. The model combines hierarchical motor representations with probabilistic forward models and unsupervisedly learned inverse models. Our evaluations with camera data of human gesturing have hitherto produced promising results with respect to a robust recognition and meaningful classification of presented gestures, making use of a growing resonant motor repertoire shared between all sensorimotor processes. The hierarchical nature of the model considers not only the mere spatio-temporal features but also more abstract levels, from the form and trajectory towards the meaning of a gesture. Using a unified motor representation for both perception and action allows direct interactions between these bottom-up and top-down processes and enables the robot to interact in more natural and socially adept ways.

Future work will further extend this line of research and tackle the symmetrical use of the resonant representations for both perceiving and generating gestures, which paves the way toward social human-robot interaction with features like mimicry and alignment. Moreover, since fingers contribute significantly in many co-verbal hand-arm gestures, the introduced finger modules in the model will be realized. Consequently, the setup needs to be extended in order to sense and perceive finger configurations as well. In this context, further training with real gesture data will be necessary to determine the learning capacity of the model.

Acknowledgements. This research is supported by the Deutsche Forschungsgemeinschaft (DFG) in the Center of Excellence in "Cognitive Interaction Technology".

References

1. Amit, R., Mataric, M.: Learning movement sequences from demonstration. In: ICDL 2002: Proceedings of the 2nd International Conference on Development and Learning, pp. 203–208 (2002)
2. Botvinick, M.M.: Hierarchical models of behavior and prefrontal function. Trends in Cognitive Sciences 12(5), 201–208 (2008), http://www.sciencedirect.com/science/article/B6VH9-4S95WHD-1/2/f02cfdf1fde7f8df4f4d2d52da7acd7b
3. Breazeal, C., Buchsbaum, D., Gray, J., Gatenby, D., Blumberg, B.: Learning from and about others: Towards using imitation to bootstrap the social understanding of others by robots. Artificial Life 11(1-2), 31–62 (2005), http://dx.doi.org/10.1162/1064546053278955
4. Calinon, S., Billard, A.: Recognition and Reproduction of Gestures using a Probabilistic Framework Combining PCA, ICA and HMM. In: 22nd International Conference on Machine Learning, pp. 105–112 (2005)
5. Dautenhahn, K.: Socially intelligent robots: Dimensions of human - robot interaction. Philosophical Transactions of the Royal Society B: Biological Sciences 362(1480), 679–704 (2007)
6. Dijksterhuis, A., Bargh, J.: The perception-behavior expressway: Automatic effects of social perception on social behavior. Advances in Experimental Social Psychology 33, 1–40 (2001)
7. Chen, F.-S., Fu, C.-M., Huang, C.L.: Hand gesture recognition using a real-time tracking method and hidden markov models. Image and Vision Computing 21, 745–758 (2003)
8. Hamilton, A., Grafton, S.: The motor hierarchy: From kinematics to goals and intentions. In: Attention and Performance, vol. 22. Oxford University Press, Oxford (2007)
9. Haruno, M., Wolpert, D.M., Kawato, M.: Mosaic model for sensorimotor learning and control. Neural Computation 13(10), 2201–2220 (2001), http://www.mitpressjournals.org/doi/abs/10.1162/089976601750541778
10. Haruno, M., Wolpert, D.M., Kawato, M.: Hierarchical mosaic for movement generation. International Congress Series 1250, 575–590 (2003), http://www.sciencedirect.com/science/article/B7581-49N7DHR-1J/2/83e9a135a8a183a9f18da5a66dcd3bbf Cognition and emotion in the brain. Selected topics of the International Symposium on Limbic and Association Cortical Systems
11. Johnson, M., Demiris, Y.: Hierarchies of coupled inverse and forward models for abstraction in robot action planning, recognition and imitation. In: Proceedings of the AISB 2005 Symposium on Imitation in Animals and Artifacts (2005)
12. Kopp, S., Graeser, O.: Imitation learning and response facilitation in embodied agents. In: Gratch, J., Young, M., Aylett, R.S., Ballin, D., Olivier, P. (eds.) IVA 2006. LNCS (LNAI), vol. 4133, pp. 28–41. Springer, Heidelberg (2006)
13. Kopp, S., Wachsmuth, I.: Synthesizing multimodal utterances for conversational agents. Journal of Computer Animation and Virtual Worlds 15(1), 39–52 (2004)
14. Kopp, S., Wachsmuth, I., Bonaiuto, J., Arbib, M.: Imitation in embodied communication – from monkey mirror neurons to artificial humans. In: Wachsmuth, I., Lenzen, M., Knoblich, G. (eds.) Embodied Communication in Humans and Machines, pp. 357–390. Oxford University Press, Oxford (2008)

15. Natalie Sebanz, G.K.: The role of the mirror system in embodied communication. In: Wachsmuth, I., Lenzen, M., Knoblich, G. (eds.) Embodied Communication in Humans and Machines, ch. 7, pp. 129–149. Oxford University Press, Oxford (2008)
16. Schutz-Bosbach, S., Prinz, W.: Perceptual resonance: action-induced modulation of perception. Journal of Trends in Cognitive Sciences 11(8), 349–355 (2007)
17. Wolpert, D.M., Doya, K., Kawato, M.: A unifying computational framework for motor control and social interaction. Philos Trans. R. Soc. Lond. B. Biol. Sci. 358(1431), 593–602 (2003), http://dx.doi.org/10.1098/rstb.2002.1238

Spatio-Temporal Situated Interaction in Ambient Assisted Living

Bernd Krieg-Brückner and Hui Shi

Abstract. This paper presents ongoing work on qualitative spatio-temporal representation and reasoning to support situated interaction in AAL environments, modelling navigation tasks, 2D/3D relations between objects, activity plans, and integration of context sensitive information into the user's intentions and the system's feedbacks.

1 Motivation

The major goal of Ambient Assisted Living (AAL) is to extend the independent and self-consistent life of elderly people, thus enhancing their quality of life and reducing their dependency on personal health care [18]. An AAL system should meet several requirements (cf. [3, 11, 15, 23]), for example, to be *personalized* to the user's needs, *adaptive* to the user's actions and environment, *anticipating* the user's desires. It is imperative that such systems enable users to interact with them in a natural way, since these users have usually little or no formal training in information technologies.

Current research on situated communication focusses on human-human (cf. [13]), human-robot (cf. [25, 12]), or artificial agents (cf. [5]). In [5] Brenner and Kruijff-Korbayová report on an approach to situated dialogue from the perspective of multi-agent planning. In their work, the dialogue between several artificial agents in a household domain is simulated. Since every agent has its capabilities and constraints, several agents need to collaborate to achieve individual goals. However, there is no particular concern for time and space.

Bernd Krieg-Brückner · Hui Shi
DFKI Bremen, Safe and Secure Cognitive Systems, and
SFB/TR8 Spatial Cognition, Universität Bremen, Germany
e-mail: {Bernd.Krieg-Brueckner,Hui.Shi}@DFKI.de

H. Ritter et al. (Eds.): Human Centered Robot Systems, COSMOS 6, pp. 205–214.
springerlink.com

Moreover, conceptual knowledge representations are used in some systems to enable human-robot interaction. [26] presents an approach to create conceptual spatial representations of indoor environments. Their model consists of several layers of map abstraction supporting human-robot dialogue. Our conceptual model, developed in [22], on the other hand, is based on qualitative spatial representation and reasoning. In addition, Route Graphs provide topological graph structures to integrate route segments into a route, or routes into Route Graphs. An advantage of using qualitative spatial calculi is that their algebraic properties and reasoning mechanisms enable automatic generation of qualitative spatial relations from metric data, cognitive representation of user's spatial knowledge and detection of spatial mismatches between users' knowledge and the robot's internal map by reasoning about spatial relations.

The main focus of our current work is the interaction between users and an AAL environment, specifically, spatial and temporal aspects. An AAL environment usually consists of a range of smart devices (e.g., climate control, light, electric doors, access control, household appliances, smart furniture), mobile assistants and portable interaction devices (e.g., the iPhone). We take the Bremen Ambient Assisted Living Lab (BAALL) and mobile assistants such as the *Rolland* wheelchair or the *iWalker* as an example [11]; these are equipped with laser range sensors, various assistants compensate for diminishing physical and cognitive faculties: the *driving assistant* avoids obstacles and helps passing through a door; the *navigation assistant* autonomously guides along a selected route; a *head joystick* and spoken natural language dialogue ease interaction. To communicate and execute an activity issued by a user, such as "go to the bed", "go to the bathroom" or "prepare a meal", interaction between the user, mobile assistants and smart devices are indispensable.

We begin in Section 2 with an introduction to an overall architecture of the interaction management system. In Section 3 we discuss the application of qualitative spatial modelling for the contextualization of users' activities; and in Section 4 an approach based on temporal representation and reasoning is presented for the collaboration between various activities in situated interaction between users and an AAL environment. We conclude with future work in Section 5.

2 An Architecture for Situated Interaction in AAL

Fig. 1 shows an overall architecture for processing situated interaction in an AAL environment. The three central components are *Context Manager*, *Interaction Manager* and *Coordination Manager*. The context manager is responsible for the integration of the context information into users' inputs or the system's outputs, whereas the *Spatial Knowledge Manager* represents the spatial environment, provides spatial functionalities, and carries out spatial reasoning; the *Activity Manager* makes an activity plan for the intention identified in the recent user input; and *Home Conception* contains the ontological definitions of the AAL environment used by several components.

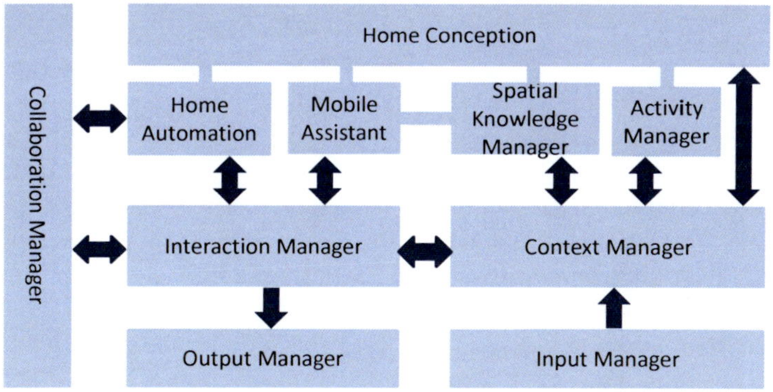

Fig. 1 The overall architecture for interaction processing

After getting an activity plan from the Context Manager, the Interaction Manager asks the Coordination Manager to check whether the plan can be carried out in the current situation. If there is no conflict with other existing activities, then the Interaction Manager sends the activity to the Home Automation or a Mobile Assistant such that it can be carried out. The Context Manager is also responsible for providing spatial context to generate the system's feedback.

Now we are going to show the interaction processing with an example in BAALL. Suppose Marion is in the dressing room, and Alfred is in the bedroom and would like iWalker to take him into the bathroom. Alfred's intention is first sent to the Context Manager. Using the Spatial Knowledge Manager and the Home Conception, the Context Manager will get iWalker's location. The Spatial Knowledge Manager plans then the route(s) from the current location to the bathroom. The following two activities are then constructed (we suppose there are two possible routes r_1 and r_2): $take(iWalker, r_1)$, $take(iWalker, r_2)$.

Accordingly, two activity plans are made by the Activity Manager; each contains the iWalker's movement following the route and activities like opening and closing doors on the route. By checking the practicability of these activities, the Collaboration Manager detects that there is a locked door on either route, since Marion has locked the doors on both sides of the dressing room for changing, thus also denying access to the bathroom. Only after these doors are unlocked, iWalker can take Alfred to the bathroom. Thus, the following interaction, for example, is supported by our system:

```
(1) Alfred:  Take me to the bathroom.
(2) iWalker: Wait a while, since Marion is in the
    dressing room.
                ......
(3) iWalker: Now I can take you to the bathroom, should I?
(4) Alfred:  Yes, please.
```

3 Spatial Knowledge in Context-Sensitive Interaction

To enable spatio-temporally aware interaction as the example in the last section shows, environment and activity modelling is necessary. The focus of the current work is twofold: contextualizing users' intentions and the system's feedbacks by spatial modelling, which is going to be discussed in this section, and collaborative interaction using temporal modelling, discussed in the next section. These provide the foundations for robust and natural interaction with an AAL environment.

Using qualitative spatial knowledge in human-robot interaction on navigation tasks have been investigated in [22], where qualitative spatial representation is used to modelling navigation functions like localization and route planning. Furthermore, qualitative spatial reasoning, based on the Double-Cross Calculus [7, 27], allows detecting spatial knowledge mismatches and generating clarification dialogues. In addition to spatial relations between locations and movements, which are essential for navigation tasks, additional spatial relations are necessary in the AAL context. Object localization, for example, needs the representation of spatial relations between objects. As empirical studies show (cf. [9]), relations in both, relative frames of reference ("left of") and global frames ("north of"), are used by people while describing object locations. An intuitive answer to the user's question "where is the sugar?" is, for example, "the sugar is on the left of the flour". A system feedback giving the x-, y- (and possibly z-) coordinates of the location in a global metric reference system would be unacceptable.

So far, we have been using the Double-Cross Calculus [7, 27], with the Route Graph, for reasoning about navigation (with mobility assistants such as the wheelchair Rolland) in the horizontal plane. When we now also consider interactions in BAALL, say with objects in a cupboard of the kitchenette; in principle, we are dealing with a 3D reference system – but not quite: the user is moving in the horizontal 2D plane to navigate to the kitchenette (is perhaps taken there automatically by Rolland); the cupboard moves down such that its lowest shelf is just below eye-level; then the interaction with the contents of the cupboard proceeds starting from the facing vertical 2D plane of the cupboard structured with its shelves; then it also considers its contents reaching into the depth of a shelf in a 2D horizontal plane. In such a structured situation, it is much more intuitive to break down the 3D relation of the flour on the facing middle shelf to the sugar on the upper shelf in the left corner way in the back (or, equivalently but worse: in the back on the left of the upper shelf) into a sequence of 2D relations: "look at the spaghetti, left on the upper shelf; you find the sugar 3 deep behind it", or, in fact, a sequence of three 1D relations, describing eye movements: "on the upper shelf, left, 3 deep behind the spaghetti"; just what is more natural will be a matter of further linguistic evaluation.

There are several qualitative models and calculi, such as the Cardinal Direction Calculus [6] or the Star Calculus [20], which are suitable for the description of the qualitative relations between object locations. Here we take the Star Calculus as an example, which specifies qualitative direction relations (e.g. east, south, west and north) between two points with respect to a given reference direction. Fig. 2 shows the Star Calculus with 4 lines, where the directions from the central point p to any

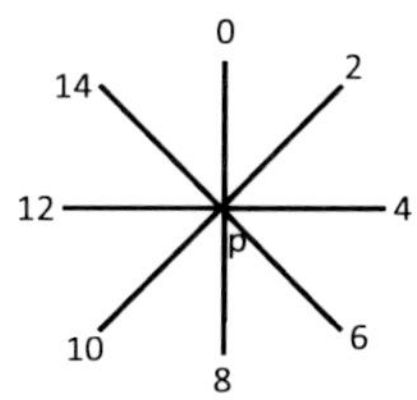

Fig. 2 The Star calculus with 4 lines

point on a half line is denoted by an even number, the sectors between two half lines are denoted by odd numbers, skipped here. A well-known application of the Star Calculus is the representation of spatial relations between geometric objects (cf. [14]).

Consider the "sugar" example again. Suppose the flour is the most recently visited object, thus we take it as the current central point, denoted as p. The reference direction is the view direction of the user, say the line 0 starting from p, denoted as 0_p. Now, the qualitative relations between the relevant objects, for example located in a cupboard, can be computed from their metric location using the reasoning mechanism of the calculus. $s \in 12_p$ represents that the sugar at location s is somewhere on the half line 12. Thus we know that the sugar is on the left of the flour from the user's perspective. However, if there is an object between the flour and the sugar, say spaghetti, then it is probably clearer to say "The sugar is on the left of the spaghetti, (which are) on the left of the flour". This requires to discover the existence of an object o that is in the direction 12_p (i.e. o is on the left of p) and in the direction 4_s (i.e. it is on the right of s). Again, the spatial reasoning of the Star Calculus helps us to solve the problem.

Our previous work [22] and the above example illustrate that qualitative spatial representation and reasoning provide necessary theories for modeling spatial knowledge of an AAL environment. In addition, a higher-level management of spatial relations is needed, such that reasoning plans can be constructed for the generation of spatially aware feedbacks.

4 Temporal Reasoning for Collaborative Interaction

In fact, in a typical AAL environment, in which parallel execution of activities is allowed, we should consider modelling at different levels. Activity Theory has been applied in various areas, such as system design (cf. [4]) or Human-Computer-Interaction (cf. [17]). As defined by Activity Theory, an activity can be broken down into actions, and actions into operations. In the AAL context, operations are direct device controls, such as opening or closing a sliding door, or turning a light on or off.

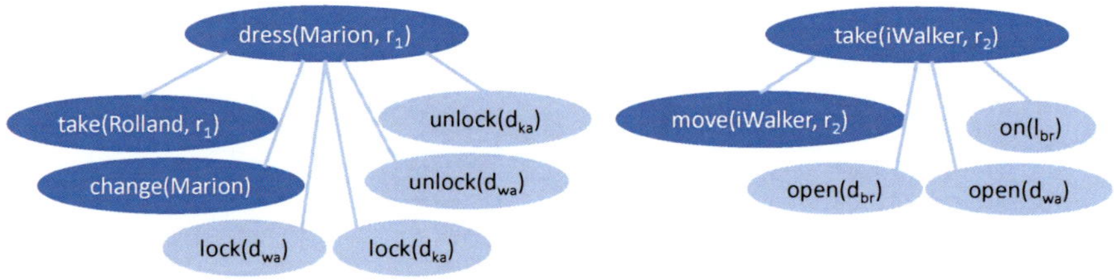

(a) Decomposition of *dress* (b) Decomposition of *take*

Fig. 3 Two sample activities

Consider the example in Section 2 again. Fig. 3 shows the decomposition of the activity $dress(Marion, r_1)$ and $take(iWalker, r_2)$ respectively. In the figure, d_{wa} and d_{ka} are the sliding doors of the dressing room to the working area (in the bedroom) and to the kitchen area, and d_{br} the sliding door of the bathroom (accessible only via the dressing room); r_1 is Rolland's route to take Marion to the dressing room, and r_2 iWalker's route to take Alfred to the bathroom. We suppose that d_{wa} and d_{br} are on the route r_2; *open*, *lock*, *unlock* are operations on doors and *on* is an operation on lights; "move" and "change" can be further divided into subactions or operations, skipped here.

Table 1 Four temporal relations from Allen's Interval Calculus

Relation	Interpretation
$A < B$	*A* takes place before *B*
$A\ s\ B$	*A* starts *B* and *B* takes longer than *A*
$A\ d\ B$	During the event *B*, *A* takes place
$A = B$	*A* and *B* start and terminate simultaneously

If (Rolland with) Marion is in the dressing room and the doors d_{wa} and d_{ka} are locked, then the operation $open(d_{wa})$ of the activity $take(iWalker, r_2)$ cannot be executed. Thus, the "take" activity cannot be carried out at the moment.

Now we are going to demonstrate the idea of temporal modelling and reasoning about relations between activities, actions and operations using Allen's Interval Calculus [2]. It defines possible relations between time intervals and provides a composition table for reasoning about temporal descriptions of events (activities). Table 1 presents four of its 13 base relations. The applications of Allen's Interval Calculus in planning and scheduling (cf. [1]) and in graphic processing (cf. [16]), for example, have been reported in the literature.

Now, the activity “take” can be represented by the set of relations S_1. It specifies that the duration of the activity “move” and “take” are the same, while operations of opening doors and turning on the light should happen during the “move” activity. Similarly, the set S_2 of relations represents the activity “dress”, where the “take” activity should take place before the “lock” operations, but they should take place before the “change” activity; the “unlock” operations can only take place after “change”, but during the main activity.

$$\begin{aligned} S_1 = \{ & \ move(iWakler, r_2) = take(iWalker, r_2), \quad open(d_{wa})\ d\ move(iWakler, r_2), \\ & \ open(d_{br})\ d\ move(iWakler, r_2), \quad on(l_{br})\ d\ move(iWakler, r_2) \ \} \end{aligned}$$

$$\begin{aligned} S_2 = \{ & \ take(Rolland, r_1)\ s\ dress(Marion, r_1), \\ & \ take(Rolland, r_1) < lock(d_{wa}), \quad take(Rolland, r_1) < lock(d_{ka}), \\ & \ lock(d_{wa}) < change(Marion), \quad lock(d_{ka}) < change(Marion), \\ & \ change(Marion) < unlock(d_{wa}), \quad change(Marion) < unlock(d_{ka}), \\ & \ unlock(d_{ka})\ d\ dress(Marion, r_1), \quad unlock(d_{wa})\ d\ dress(Marion, r_1) \} \end{aligned}$$

To do situated temporal reasoning, the Collaboration Manager should maintain a set of temporal relations which are situation dependent. Therefore, we introduce the class of *situated temporal relations*. A situated temporal relation is a temporal relation with a condition and interpreted as: if the condition is true, then the temporal relation should hold. For example:

$$isLocked(d_{wa}) \Rightarrow unlock(d_{wa}) < open(d_{wa})$$

means that if the sliding door of the working area is locked, the operation “unlock” should take place before “open”. Situated temporal relations are activity independent, and may introduce additional relations between activities according to the environment situation. In our example, Marion is already in the dressing room, changing, i.e. the following actions and operations have already been carried out: $take(Rolland, r_1)$, $lock(d_{wa})$ and $lock(d_{ka})$. Thus, the condition $isLocked(d_{wa})$ is true in the current situation and the operation $open(d_{wa})$ of the $take(iWakler, r_2)$ can only take place after the operation $unlock(d_{wa})$. Since $unlock(d_{wa})$ is an operation of the activity $dress(Marion, r_1)$, the interaction manager concludes that the activity $take(iWakler, r_2)$ should wait for $dress(Marion, r_1)$. As a result the Output Manager generates the feedback “Please wait a while, since Marion is in the dressing room”. After the execution of $unlock(d_{wa})$, the condition $isLocked(d_{wa})$ becomes false. The operation $open(d_{wa})$ can then take place, and the activity $take(iWakler, r_2)$ is allowed to start. As a consequence, iWalker asks: “Now I can take you to the bathroom, should I?”.

As stated in the previous and current section, to enable users to interact with an AAL environment naturally, we introduced models of both spatial and temporal relations. Consequently, the integration of these models should be considered as well. There are several researches in this direction, for example, [8] presents a calculus that combines RCC-8 with Allen’s Interval Calculus, while [19] introduces a model that combines the Cardinal Direction Calculus with Allen’s Interval Calculus.

5 Conclusion

In this paper we showed examples for the application of qualitative spatio-temporal representation and reasoning in situated interaction in AAL. Specifically, spatial representation and reasoning based on the Double-Cross calculus is used for human-robot interaction on navigation tasks; the Star Calculus is applied to model relations between objects for solving object localization problems; and we use Allen's Interval calculus to specify activity plans and relations between activities, actions and operations. Spatio-temporal reasoning then allows the system to integrate context sensitive information into the user's intentions and the system's feedbacks. Thus, situated interaction is achieved.

Several components developed in our research center, for example, the spatio-temporal reasoning toolbox *SparQ* [24], the formal method based framework for the development of interaction controls *SimSpace* [10] and the generic dialogue system *DAISIE* [21], provide foundations for the development of the interaction system described in this paper. However, further development of various management components (Fig. 1) is yet ongoing work. The integration of more spatio-temporal calculi will be investigated in the future. Moreover, we will further evaluate the interaction system with real users for its acceptance and reliability.

Acknowledgements. We gratefully acknowledge the support of the Deutsche Forschungsgemeinschaft (DFG) through the Collaborative Research Center SFB/TR8 - Project I3-[SharC] "Shared Control via Interaction", and the EU in project SHARE-it (FP6-045088).

References

1. Allen, J., Koomen, J.: Planning using a temporal world model. In: Proceedings of the Eighth International Joint Conference on Artificial Intelligence (1983)
2. Allen, J.F.: Maintaining knowledge about temporal intervals. CACM 26(11), 832–843 (1983)
3. Augusto, J.C., McCullagh, P.: Ambient intelligence: Concepts and applications. Computer Science and Information Systems 4(1), 228–250 (2007)
4. Bardram, J.E.: Scenario-based design of cooperative systems. In: Proceedings of COOP 1998 (1998)
5. Brenner, M., Kruijff-Korbayová, I.: A continual multiagent planning approach to situated dialogue. In: Proceedings of the 12th Workshop on the Semantics and Pragmatics of Dialogue (2008)
6. Frank, A.U.: Qualitative spatial reasoning with cardinal directions. In: Proceedings of the Seventh Austrian Conference on Artificial Intelligence. Springer, Heidelberg (1991)
7. Freksa, C.: Using orientation information for qualitative spatial reasoning. In: Frank, A.U., Formentini, U., Campari, I. (eds.) GIS 1992. LNCS, vol. 639, pp. 162–178. Springer, Heidelberg (1992)

8. Gerevini, A., Nebel, B.: Qualitative spatio-temporal reasoning with rcc-8 and allen's interval calculus: Computational complexity. In: Proceedings of ECAI 2002, pp. 312–316 (2002)
9. Hois, J., Tenbrink, T., Ross, R., Bateman, J.: The Generalized Upper Model spatial extension: a linguistically-motivated ontology for the semantics of spatial language. SFB/TR8 internal report, Collaborative Research Center for Spatial Cognition, University of Bremen, Germany (2008)
10. Jian, C., Shi, H., Krieg-Brückner, B.: Simspace: A tool to interpret route instructions with qualitative spatial knowledge. In: Technical Report of the AAAI Spring Symposium on Benchmarking of Qualitative Spatial and Temporal Reasoning Systems (2009)
11. Krieg-Brückner, B., Röfer, T.T., Shi, H., Gersdorf, B.: Mobility Assistance in the Bremen Ambient Assisted Living Lab. Journal of Gerontology. Special Section: Technology and Aging: Integrating Psychological, Medical, and Engineering Perspectives (to appear)
12. Kruijff, G.J.M., Zender, H., Jensfelt, P., Christensen, H.I.: Clarification dialogues in human-augmented mapping. In: Proceedings of the 1st ACM SIGCHI/SIGART Conference on Human-Robot Interaction (2006)
13. Lücking, A., Rieser, H., Staudacher, M.: SDRT and multi-modal situated communication. In: Proceedings of the 10th Workshop on the Semantics and Pragmatics of Dialogue, pp. 72–79 (2006)
14. Mitra, D.: Representing geometrical objects by relational spatial constraints. In: Proceedings of Knowledge-based Computer Systems, KBCS (2002)
15. Mukasa, K.S., Holzinger, A., Karshmer, A.: Intelligent user interfaces for ambient assisted living. In: Proceedings of the First International Workshop IUI4AAL 2008 (2008)
16. Nabil, M., Shepherd, J., Ngu, A.H.H.: 2D Projection Interval Relationships: A Symbolic Representation of Spatial Relationships. In: Egenhofer, M.J., Herring, J.R. (eds.) SSD 1995. LNCS, vol. 951. Springer, Heidelberg (1995)
17. Nardi, B.A.: Context and Consciousness: Activity Theory and Human-Computer Interaction. MIT Press, Cambridge (1996)
18. Nehmer, J., Karshmer, A., Becker, M., Lamm, R.: Living Assistance Systems - an Ambient Intelligence Approach. In: Proceedings of the 28th International Conference on Software Engineering (2006)
19. Ragni, M., Wölfl, S.: Temporalizing Cardinal Directions: From Constraint Satisfaction to Planning. In: Proceedings of the Tenth International Conference on Principles of Knowledge Representation and Reasoning, pp. 472–480. AAAI Press, Menlo Park (2006)
20. Renz, J., Mitra, D.: Qualitative direction calculi with arbitrary granularity. In: Zhang, C., Guesgen, H.W., Yeap, W.-K. (eds.) PRICAI 2004. LNCS (LNAI), vol. 3157, pp. 65–74. Springer, Heidelberg (2004)
21. Ross, R.J.: Situated Dialogue Systems: Agency & Spatial Meaning in Task-Oriented Dialogue. Ph.D. thesis, Universitat Bremen (2009)
22. Shi, H., Krieg-Brückner, B.: Modelling Route Instructions for Robust Human-Robot Interaction on Navigation Tasks. International Journal of Software and Informatics 2(1), 33–60 (2008)
23. Stefanov, D.H., Bien, Z., Bang, W.C.: The Smart House for Older Persons and Persons with Physical Disabilities: Structure, Technology arrangements, and Perspectives. IEEE Transactions on Neural Systems and Rehabilitation Engineering 12(2) (2004)
24. Wallgrün, J.O., Frommberger, L., Wolter, D., Dylla, F., Freksa, C.: A toolbox for qualitative spatial representation and reasoning. In: Barkowsky, T., Knauff, M., Ligozat, G., Montello, D.R. (eds.) Spatial Cognition 2007. LNCS (LNAI), vol. 4387, pp. 39–58. Springer, Heidelberg (2007)

25. Zender, H., Jensfelt, P., Mozos, O.M., Kruijff, G.J.M., Burgard, M.: An integrated robotic systen for spatial understanding and situated interaction in indoor environments. In: Proc. of the Conference on Artificial Intelligence (2007)
26. Zender, H., Mozos, O.M., Jensfelt, P., Kruijff, G.J.M., Burgard, M.: Conceptual spatial representations for indoor mobile robots. Robotics and Autonomous Systems 56 (2008)
27. Zimmermann, K., Freksa, C.: Qualitative spatial reasoning using orientation, distance, and path knowledge. Applied Intelligence 6, 49–58 (1996)

Author Index

Cognitive Systems Monographs

Edited by R. Dillmann, Y. Nakamura, S. Schaal and D. Vernon

Vol. 1: Arena, P.; Patanè, L.: (Eds.)
Spatial Temporal Patterns for
Action-Oriented Perception
in Roving Robots
425 p. 2009 [978-3-540-88463-7]

Vol. 2: Ivancevic, T.T.; Jovanovic, B.;
Djukic, S.; Djukic, M.; Markovic, S.:
Complex Sports Biodynamics
326 p. 2009 [978-3-540-89970-9]

Vol. 3: Magnani, L.:
Abductive Cognition
534 p. 2009 [978-3-642-03630-9]

Vol. 4: Azad, P.:
Cognitive Systems Monographs
270 p. 2009 [978-3-642-04228-7]

Vol. 5: de de Aguiar, E.:
Animation and Performance Capture
Using Digitized Models
xxx p. 2009 [978-3-642-10315-5]

Vol. 6: Ritter, H.; Sagerer, G.;
Dillmann, R.; Buss, M.:
Human Centered Robot Systems
216 p. 2009 [978-3-642-10402-2]